THE MATHEMATICS OF ORIGAMI

When you see a paper crane, what do you think of? A symbol of hope, a delicate craft, The Karate Kid? What you might not see, but is ever present, is the fascinating mathematics underlying it. Origami is increasingly applied to engineering problems, including origami-based stents, deployment of solar arrays in space, architecture, and even furniture design. The topic is actively developing, with recent discoveries at the frontier (e.g., in rigid origami and in curved-crease origami) and an infusion of techniques and algorithms from theoretical computer science. The mathematics is often advanced, but this book instead relies on geometric intuition, making it accessible to readers with only a high school geometry and trigonometry background. Through careful exposition, more than 160 color figures, and 49 exercises all completely solved in an Appendix, the beautiful mathematics leading to stunning origami designs can be appreciated by students, teachers, engineers, and artists alike.

Joseph O'Rourke is Olin Professor Emeritus of Computer Science at Smith College where he has a joint appointment in Mathematics. He has written or coauthored eight books, including two textbooks and two books written for high school students: *How to Fold It* (2011) and *Pop-Up Geometry* (2022). His research is in computational geometry, developing algorithms for geometric computations, and he has published more than 175 papers in journals and conference proceedings in this area. He has won several awards, including a Guggenheim Fellowship in 1987 and the NSF Director's Award for Distinguished Teaching Scholars in 2001. He was named an ACM Fellow in 2012.

The Mathematics of Origami

Joseph O'Rourke
Smith College, Massachusetts

Shaftesbury Road, Cambridge CB2 8EA, United Kingdom

One Liberty Plaza, 20th Floor, New York, NY 10006, USA

477 Williamstown Road, Port Melbourne, VIC 3207, Australia

314–321, 3rd Floor, Plot 3, Splendor Forum, Jasola District Centre, New Delhi – 110025, India

103 Penang Road, #05–06/07, Visioncrest Commercial, Singapore 238467

Cambridge University Press is part of Cambridge University Press & Assessment, a department of the University of Cambridge.

We share the University's mission to contribute to society through the pursuit of education, learning and research at the highest international levels of excellence.

www.cambridge.org
Information on this title: www.cambridge.org/9781009687355

DOI: 10.1017/9781009687362

© Joseph O'Rourke 2026

This publication is in copyright. Subject to statutory exception and to the provisions of relevant collective licensing agreements, no reproduction of any part may take place without the written permission of Cambridge University Press & Assessment.

When citing this work, please include a reference to the DOI 10.1017/9781009687362

First published 2026

A catalogue record for this publication is available from the British Library

A Cataloging-in-Publication data record for this book is available from the Library of Congress

ISBN 978-1-009-68735-5 Hardback
ISBN 978-1-009-68738-6 Paperback

Cambridge University Press & Assessment has no responsibility for the persistence or accuracy of URLs for external or third-party internet websites referred to in this publication and does not guarantee that any content on such websites is, or will remain, accurate or appropriate.

For EU product safety concerns, contact us at Calle de José Abascal, 56, 1°, 28003 Madrid, Spain, or email eugpsr@cambridge.org.

Contents

Preface *page* ix
    1   Mechanics and Conventions   x
    2   Chapter Summaries   xi

1 Introduction   1
    1.1   Mathematics Boxes   1
    1.2   Creases, M/V   1
    1.3   Dihedral vs. Fold Angles   3
    1.4   Why Theoretical Computer Science?   5
    1.5   Why Theorems and Proofs?   5

2 Stamp Folding   6
    2.1   Introduction   6
    2.2   Exponential Growth   7
    2.3   Dragon Curve   8
    2.4   Counting Stamp Foldings   10
    2.5   Map Folding   14
    2.6   Technical Notes   18

3 Flat Vertex Folds   19
    3.1   Introduction   19
    3.2   Maekawa's Theorem   21
    3.3   Even-Degree Lemma   24
    3.4   Local-Min Lemma   25
    3.5   Kawasaki's Theorem   26
    3.6   Technical Notes   32

4 Flat Folding Is Hard   33
    4.1   Introduction   33
    4.2   P, NP-Complete, NP-Hard   34
    4.3   Flat Folding Is NP-Hard: Proof Sketch   38
    4.4   Turing-Completeness   46
    4.5   Flat Folding Is Turing-Complete: Proof Sketch   51
    4.6   Technical Notes   58

5	**Rigid Origami and Degree-4 Vertices**	59
	5.1 Introduction	59
	5.2 Rigid Origami	59
	5.3 Miura Map Fold	64
	5.4 The Square Twist	77
	5.5 Octahedron Top	84
	5.6 Rigidly Flat-Foldable Is NP-Hard	85
	5.7 Technical Notes	86
6	**Origami Design**	87
	6.1 Introduction	87
	6.2 The Design Process	87
	6.3 Uniaxial Bases	89
	6.4 Circle/River Method	99
	6.5 Straight Skeleton	102
	6.6 Technical Notes	104
7	**Fold & 1-Cut**	105
	7.1 History: Harry Houdini	105
	7.2 Theorem Statement	105
	7.3 Straight-Skeleton Proof	108
	7.4 Disk-Packing Proof	117
	7.5 Flattening Polyhedra	119
	7.6 Technical Notes	120
8	**Curved-Crease Origami**	121
	8.1 Introduction	121
	8.2 *Vesica Piscis*	122
	8.3 Pleated Hyperbolic Paraboloid	125
	8.4 Curvature	128
	8.5 Three General Properties	131
	8.6 Rigidly Foldable Curved Creases	133
	8.7 Technical Notes	140
9	**Self-Folding Origami**	141
	9.1 Introduction	141
	9.2 Waterbomb Tubes	141
	9.3 Self-Folding Polyhedral Containers	144
	9.4 Paths in Configuration Space	147
	9.5 Technical Notes	152
10	**ORIGAMIZER**	153
	10.1 Introduction	153
	10.2 Strip Algorithm	154
	10.3 ORIGAMIZER Algorithm	158
	10.4 Technical Notes	167

Appendix A Beyond: Topics Not Covered		168
Appendix B Solutions to Exercises		169
B.1	Chapter 1 Exercises	169
B.2	Chapter 2 Exercises	169
B.3	Chapter 3 Exercises	170
B.4	Chapter 4 Exercises	171
B.5	Chapter 5 Exercises	174
B.6	Chapter 6 Exercises	177
B.7	Chapter 7 Exercises	178
B.8	Chapter 8 Exercises	180
B.9	Chapter 9 Exercises	181
B.10	Chapter 10 Exercises	183
References		186
Index		190

Preface

There has been a surge of interest in the mathematics of origami. On the one hand, mathematics explains the structure and dynamics of known origami models, for example, the Miura map fold (Section 5.3). But more exciting is that mathematics has led to intricate and beautiful origami constructions, for example, Robert Lang's visually stunning designs. In fact it was Lang's display at a 1996 conference of an intricate origami deer he folded using mathematical techniques—see ahead to Figure 6.1—that instantly captivated me on the topic. This burgeoning interest is further fueled by the infusion of computer science questions and techniques applied to origami, specifically algorithms and "computational complexity" results.

Since Lang's talk, several treatises have been published:

- 2003: Lang's influential *Origami Design Secrets: Mathematical Methods for an Ancient Art* (Lang 2003); second edition (Lang 2012). 758 pages.
- 2007: My book with Erik Demaine, *Geometric Folding Algorithms: Linkages, Origami, Polyhedra* (Demaine and O'Rourke 2007). 472 pages.
- 2017: Robert Lang's *Twists, Tilings, and Tessellations: Mathematical Methods for Geometric Origami* (Lang 2017). 736 pages.
- 2020: Ryuhei Uehara's *Introduction to Computational Origami* (Uehara 2020). 220 pages.
- 2020: Thomas Hull's *Origametry: Mathematical Methods in Paper Folding* (Hull 2020).[1] 332 pages.

I think it is fair to say that all of these books assume experience and knowledge of mathematics equivalent to upper-level university mathematics majors or graduate students, and professional researchers. Their combined total of 2518 (!) pages contains a wealth of beautiful mathematics, but much of it out of reach without the prerequisites, even to the most enthusiastic novice.

My goal is to nevertheless explain the advanced mathematics assuming only the equivalent of high-school mathematics (in the USA), or first-year college mathematics. In particular, calculus is not needed, but familiarity with basic geometry and trigonometry is assumed.

---

[1] I reviewed this book here: O'Rourke (2024).

The more challenging prerequisite is the mathematical "maturity" sufficient to understand proofs. Familiarity with some notion of proof beyond two-column geometry proofs is needed. I try to help develop this maturity throughout the book. To explain the advanced mathematics, I resort to proof sketches that attempt to convey the spirit without pretending to be rigorous. Proofs presented often rely more on geometric intuition than on technical arguments.

I should make clear that, unlike Lang's books, no origami folding instructions are included. Relevant templates are available at the book's website,[2] but this is not a book on how to fold specific origami models.

The remainder of this Preface will run through some of the conventions used throughout the book, followed by a concise summary of each chapter.

1 Mechanics and Conventions

References References to literature are cited as in the previous list of books—author names and (year)—referring to the References at the end of the book. To avoid the interruptions typical of an academic reference work, I decided not to cite authors and literature within the body of each chapter, instead gathering all references in *Technical Notes* at the end of each chapter.

History Almost every topic in the book has a back story, but I rely on the books written previously and their references to map out the history fairly. I remain responsible for any author slights.

Exercises There are 49 exercises, marked in three levels: [Practice] (18), [Understanding] (23), or [Challenge] (8). All exercises are solved in the final chapter, so serve in some sense as extensions of the text.

Mathematics Boxes When the exposition depends on a bit of mathematics that might be unfamiliar or overly technical, I explain it in a clearly marked (skippable) box.

Theorems and Lemmas and Open Problems Throughout, when some relatively major insight is established (or sketched, or just cited), the "theorem" is boxed. Similarly for "lemmas"—usually results underpinning theorems. Occasionally I highlight "open problems," questions at the frontier of mathematics that have so far resisted resolution.

Presentation Style Mathematicians have developed a style of introducing labels "in-line" that I embrace throughout. Here is an example from Chapter 8 (Section 8.4):

"Let p be the highest point of the curve $\mathcal{C}$, ..."

This means that in the discussion to follow, we will use p to designate that highest point, and $\mathcal{C}$ to designate the curve. Beyond the immediate discussion, it may be that p or $\mathcal{C}$ are redefined to mean different objects, but it should be

---

[2] http://cs.smith.edu/~jorourke/MathOrigami/.

clear from the context. Those with programming skills will recognize this as declaring the scope of a variable.

A similar "in-line" convention is to mark terms being defined "on the fly" via a special font. An example from Chapter 6 (Section 6.2.1):

"In a ***metric tree***, each edge has a length."

In contrast, quoted terms are often left only informally defined.

Animations I have made available more than a dozen 3D animations at http://cs.smith.edu/~jorourke/MathOrigami/. I consider these essential for understanding model dynamics, especially in Chapter 5 on rigid origami. Each animation is cited where relevant.

2 Chapter Summaries

Here follow concise summaries of the chapter contents. Although there is a natural progression, extensive inter-chapter referencing permits exploring the topics in any order.

Chapter 1: Introduction. Conventions and terminology: creases, M- and V-folds, dihedral vs. fold angles.

Chapter 2: Stamp Folding. Folding a strip of stamps leads to exponential growth and the dragon curve.

Chapter 3: Flat Vertex Folds. Folding a single vertex flat. Several constraints lead to a complete characterization, Kawasaki's Theorem 3.2.

Chapter 4: Flat Folding Is Hard. Folding multiple vertices flat is technically intractable. Its complexity leads to Turing machines and the game of LIFE.

Chapter 5: Rigid Origami and Degree-4 Vertices. Creases as hinges between rigid plates lead to dynamics determined by a beautiful half-tangent equation, with diverse practical applications.

Chapter 6: Origami Design. How to work backwards from a desired 3D model to a Mountain/Valley crease pattern, largely following Lang's uniaxial bases approach.

Chapter 7: Fold & 1-Cut. The remarkable Fold & 1-Cut Theorem 7.1, explained via straight skeletons and disk packings.

Chapter 8: Curved-Crease Origami. Explores aspects of creasing along curves, a relatively new pursuit arising primarily from art.

Chapter 9: Self-Folding Origami. Another relatively new pursuit, this time emerging from a wide variety of engineering applications.

Chapter 10: ORIGAMIZER. A high-level description of the stunning ORIGAMIZER algorithm, which essentially guides folding paper to any desired shape—technically, to any polyhedral manifold.

The one-page Appendix A lists topics that could fit under the book's title but are not included, and Appendix B gathers all the exercise solutions.

Acknowledgments Almost everything I know about the mathematics of origami I learned from Erik Demaine, Thomas Hull, Robert Lang, Klara Mundilova, Tomohiro Tachi, and Ryuhei Uehara.

It's a pleasure to acknowledge (for the second time!) my editor Kaitlin Leach's wise guidance, and to thank the entire Cambridge team for their cordial professionalism.

1

Introduction

This introductory chapter does not study any particular origami techniques or models, but establishes some conventions and terminology necessary for the remainder of the book.

1.1 Mathematics Boxes

As mentioned in the Preface, mathematics terminology and material that may not be familiar is boxed to make it easy to skip if not needed. An example is Box 1.1.

> **Box 1.1 Real Numbers**
>
> Mathematicians use the symbol $\mathbb{R}$ to designate the set of **real numbers**, numbers that can be represented by a (possibly infinite) decimal expansion. For example, $\pi = 3.1415926535\ldots$. Reals are distinguished from "natural numbers" or "whole numbers" $0, 1, 2, 3, \ldots$, and distinguished from **integers**, natural numbers plus their negatives. $\mathbb{R}^2$ represents a 2-dimensional plane, say, the xy-plane: Two real coordinates specify a point p on the plane. Similarly, $\mathbb{R}^3$ represents 3-dimensional space, often called **3-space**.

Sometimes just a footnote instead of a box suffices to explain terminology. Rather than filling this chapter with all the preliminary mathematics boxes, the boxes instead will be located where first needed, with back referencing for reminders in later chapters.

1.2 Creases, M/V

All origami in this book is created by folding a single piece of paper. No *modular origami* is included. A *crease* on a piece of paper is a line segment,

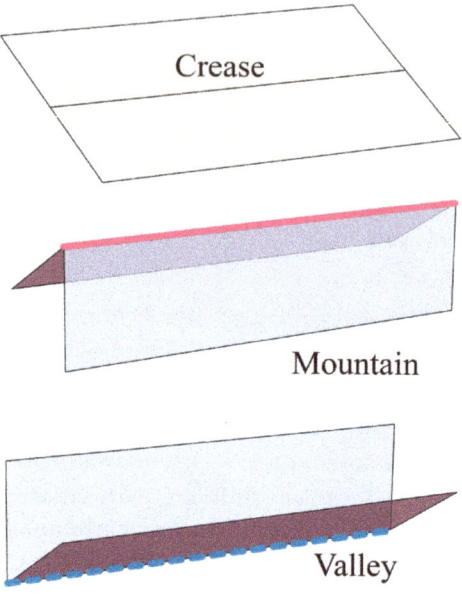

Figure 1.1 Crease and M/V folds.

which might not cross from boundary-to-boundary, but rather start and end in the interior of the paper. These internal creases are one way in which our emphasis in this book differs from the practice of origami folding, which almost always includes preparatory creases that are ultimately flattened in the final model. A *crease pattern* is the collection of creases on the paper that become specific folds in the final model.

The insistence that crease segments are straight is relaxed in Chapter 8, where we investigate curved creases.

A crease is flat. An **M/V assignment** to a crease is an indication that the crease is to be folded either as a mountain (M) or as a valley (V); see Figure 1.1. Of course a mountain fold is a valley fold viewed from the other side. Although we are defining a crease as flat, it will often be convenient to use "mountain crease" as shorthand for an M-fold crease, and similarly for V-folds. Note that an M- or V-fold does not specify the fold angle, just that the angle is either plus or minus.

How to mark figures with M/V creases in the age of color printing has not been entirely settled in the community. I use the convention that solid red = M, whereas dashed blue = V. When it seems infeasible to dash, solid blue = V.

The next section addresses a bit of a terminological conundrum.

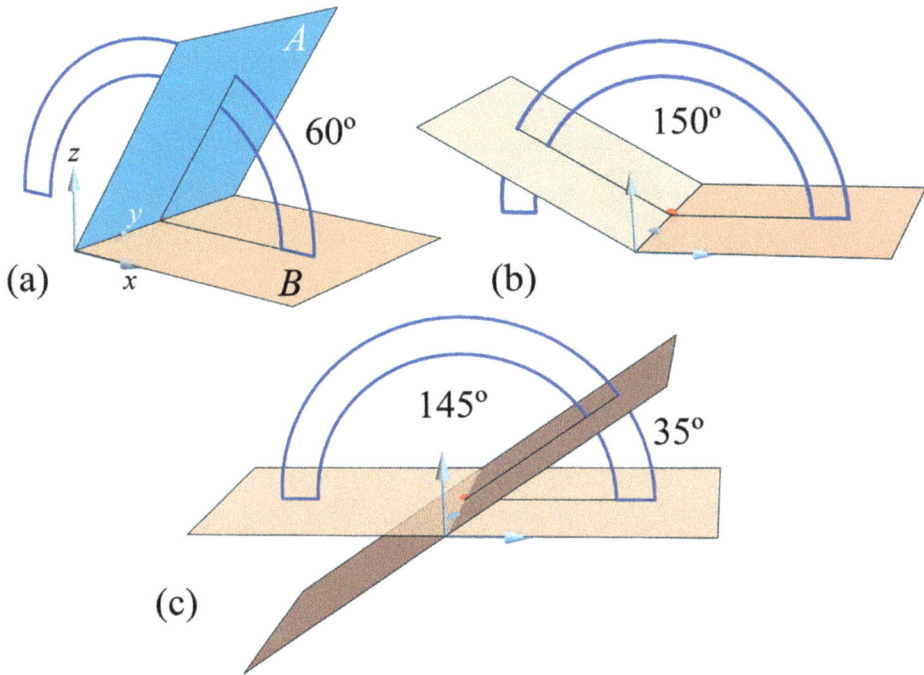

Figure 1.2 (a,b) Dihedral angles as measured by a protractor orthogonal to L, which is here the y-axis. (c) Two intersecting planes determine two supplementary angles: $35° + 145° = 180°$.

1.3 Dihedral vs. Fold Angles

Angles play a central role in the mathematics of origami. Two-dimensional angles between creases need no further explanation, but less familiar are angles between planes in 3-space, to which we now turn.

The **dihedral angle** δ is the angle between two planes: dihedral = "two sides." Any pair of nonparallel planes A and B intersect in a line $L = A \cap B$.[1] Placing a protractor in a plane orthogonal to L measures the dihedral angle δ between A and B.

Note the convention, extrapolated from Figure 1.2(a,b), that $\delta = 180°$ when A makes a flat angle with B along L (the y-axis in the figure), and $\delta = 0°$ when they make a sharp crease along L. Of course a pair of planes determine two supplementary angles at L, as indicated in Figure 1.2(c).

---

[1] $\cap$ is the symbol meaning "intersect." $\cup$ means "union."

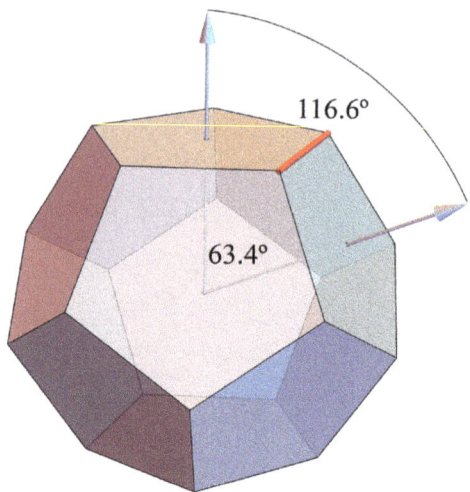

Figure 1.3 $\delta = \arccos(-1/\sqrt{5}) \approx 116.565°$. The supplementary angle ϕ is equal to the spread of the depicted normal vectors.

It is often more convenient in origami to use the supplementary **fold angle** ϕ, the angle needed to fold two planar faces to obtain a desired angle along a crease line L. Then, because $\phi + \delta = 180°$, $\phi = 0°$ when there is no fold at L, and $\phi = 180°$ when the faces are folded into a sharp crease, with the two faces overlapping to the same side of L.

Box 1.2 Normal Vectors

A **vector** is an oriented line segment, an arrow pointing from tail to head. A vector orthogonal (perpendicular) to a plane is called a **normal vector**. "Orthogonal" and "perpendicular" are synonyms, but we'll reserve "normal" for vectors.

The term "dihedral" originates in the ancient study of polyhedra. For example, the dihedral angle at each edge of a dodecahedron, the fourth of the five Platonic solids, is about 116.6°. The supplementary angle, 63.4°, can be viewed as the spread of the normal vectors (Box 1.2) to the faces sharing an edge: see Figure 1.3.

Both dihedral angles and fold angles can be positive or negative. For example, a positive fold angle is usually a mountain and a negative angle a valley. But in many contexts, only the magnitude of a dihedral or fold angle is relevant. Because "dihedral angle" is prominent in geometry and "fold angle" more common in the origami literature, it is best for readers to become comfortable with both. So we'll use whichever seems most appropriate in each particular context.

1.4 Why Theoretical Computer Science?

Why is the infusion from computer science relevant to this book's focus on the mathematics of origami?

First, theoretical computer science is arguably a part of mathematics. After all, one of the most important unsolved problem in all of mathematics is the P =? NP Millennium Prize question (see Chapter 4), which falls squarely in theoretical computer science. But second, understanding the transition from "easy" (e.g., characterization of single-vertex flat-foldability) to NP-hard (multiple-vertex flat-foldability) both explains historical difficulties and suggests feasible future foci. Computer science has illuminated aspects of origami mathematics that were previously inaccessible.

Finally, the design of NP-hard proof origami "gadgets" is both structurally revealing and—Fun!

1.5 Why Theorems and Proofs?

Why is there so much emphasis in this book on theorems and proofs?

Theorems are timeless nuggets of truth that pin-down our advance in understanding. Kawasaki's Theorem 3.2 settled once and for all time the conditions for when a single vertex can fold flat. The Degree-4 Folding Theorem 5.2 more recently uncovered a beautiful understanding of flat-foldable degree-4 vertices. Each such theorem is a step on a route to the frontier of mathematical understanding.

Theorems are established through **proofs**, watertight logical arguments that are convincing to anyone who has the background (the "mathematical maturity") to follow the arguments.

Recent advances suggest we are nowhere near resolving all the questions at the mathematical origami frontier, let alone those open problems possibly well beyond, several of which are highlighted in the following chapters. Anticipating advances, this book may need a second edition!

2

Stamp Folding

2.1 Introduction

Folding a strip of stamps into a stack is among the simplest tasks that meets the Japanese meaning of "origami": ori-"fold" -gami "paper." Despite the simplicity, there is rich mathematics hidden in these foldings. They highlight the distinction between exponential growth in contrast to polynomial growth (Section 2.2), concepts we need throughout the book. Section 2.3 reveals a surprising connection between stamp foldings and the fractal "dragon curve." We next turn to counting stamp foldings (Section 2.4)—in a branch of mathematics called *combinatorics*. We close the chapter with map folding (Section 2.5), a topic we revisit in Chapter 5 (Section 5.3).

In previous centuries, postage stamps were often sold in coiled strips, each stamp connected to the next by perforations for ease of separation, Figure 2.1 shows a short strip. Although coiled strips are still available, more commonly today stamps are sold in sheets, which we will explore as "map foldings" in Section 2.5. The prior sections will concentrate on folding a strip of identical

Figure 2.1 Strip of postage stamps.

unit-square stamps into a 1×1 stack. Of course this is easily accomplished, but despite its simplicity, it leads to fascinating and intricate mathematics.

2.2 Exponential Growth

A strip of n unit-square stamps can be folded into a single square stack by making $n-1$ folds. This can be achieved, for example, with a **pleat fold**, also known as an **accordion** fold—alternating mountain and valley folds, as in Figure 2.2.

Although the $n-1$ creases can be folded one at a time, there is a faster way to fold the strip to a single unit square. For $n = 8$, fold the whole strip in half, forming a 4×1 double-layer strip, then fold in half again, resulting in a 2×1 four-layer strip, and finally one more fold. So rather than 7 folds, the stacking is achieved with 3 folds. If n is a power of 2, $n = 2^k$, then k folds achieve the stacking, in our example, $n = 2^3$ and $k = \log_2(2^k) = 3$. See Boxes 2.1 and 2.2.

> **Box 2.1 Exponential vs. Polynomial**
>
> Generally n is used to stand for a positive integer that represents the size of a problem. For stamp folding, n is the number of stamps in a strip.
>
> **Exponential growth** refers to a function c^n, where c is some positive real number constant greater than 1, constant in the sense that c does not depend on n. Typically $c = 2$, although later we'll see $c = 3.3$.
>
> **Polynomial growth** describes the function n^k, where k is a positive number, generally an integer. An important fact for analyzing complexity is that an exponential function always eventually exceeds a polynomial function, regardless of how large is the polynomial exponent k, "eventually" in the sense that for sufficiently large n, $2^n > n^k$. This is illustrated in Figure 2.3.

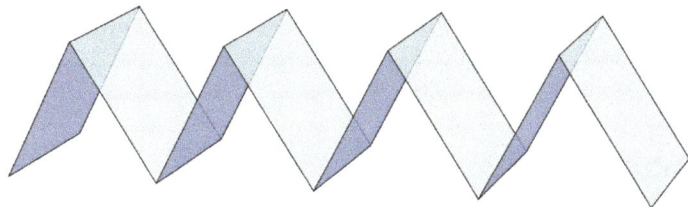

Figure 2.2 Pleat folding a strip of $n = 8$ stamps, using $n-1 = 7$ creases.

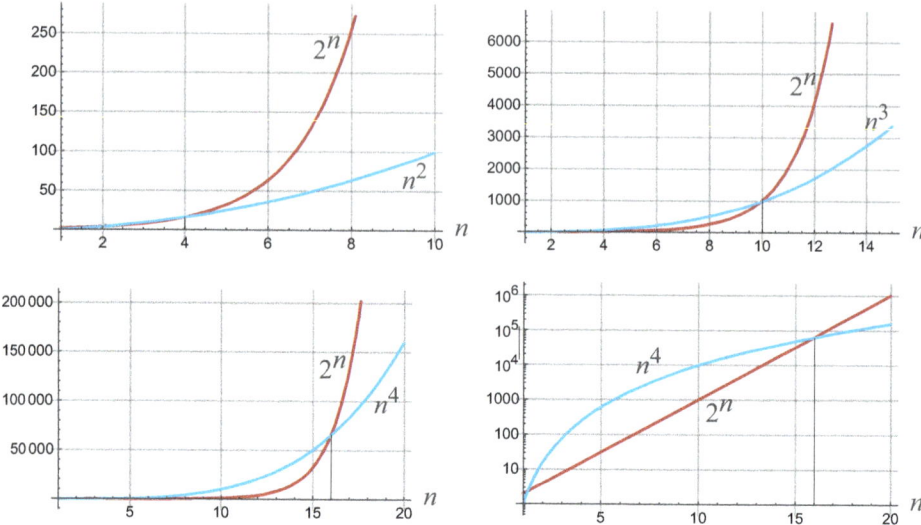

Figure 2.3 2^n compared to n^2, n^3, n^4. Crossover occurs at $n = 4, \approx 10, 16$ respectively. (The symbol $\approx$ means approximately equal.)

> **Box 2.2 Logarithms**
>
> The "log base 2" of some number n, symbolically $\log_2(n)$, is the power of 2 needed to reach n. If n is itself a power of 2, $n = 2^k$, then $\log_2(n)$ is the number of times n can be cut in half before reaching 1. So if $n = 16 = 2^4$, n can be cut in half 4 times:
>
> $$2^4 = 16 \to 8 \to 4 \to 2 \to 1.$$
>
> The fourth graph in Figure 2.3 illustrates a ***log plot***, which rescales the vertical axis so that an exponential function plots as a straight line.

> **Exercise 2.1 [Challenge] Exponential/Polynomial**
>
> Calculate at what value (approximately) of n does 2^n cross over n^{100}.

2.3 Dragon Curve

The simple repeated halving of an $n \times 1$ strip leads to the rich mathematics of the so-called ***dragon curve***. Let's start with $n = 4$. After folding in half once, and then a second time, display the stamps in a side view, with all creases 90°, so not yet folded all the way to flatness. This is shown in Figure 2.4(a), with

2.3. Dragon Curve

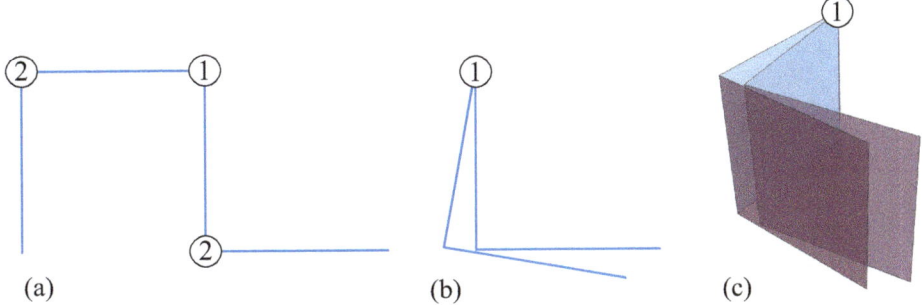

Figure 2.4 (a) Dragon curve of order-2 with crease numbers marked. (b,c) After nearly closing the first crease.

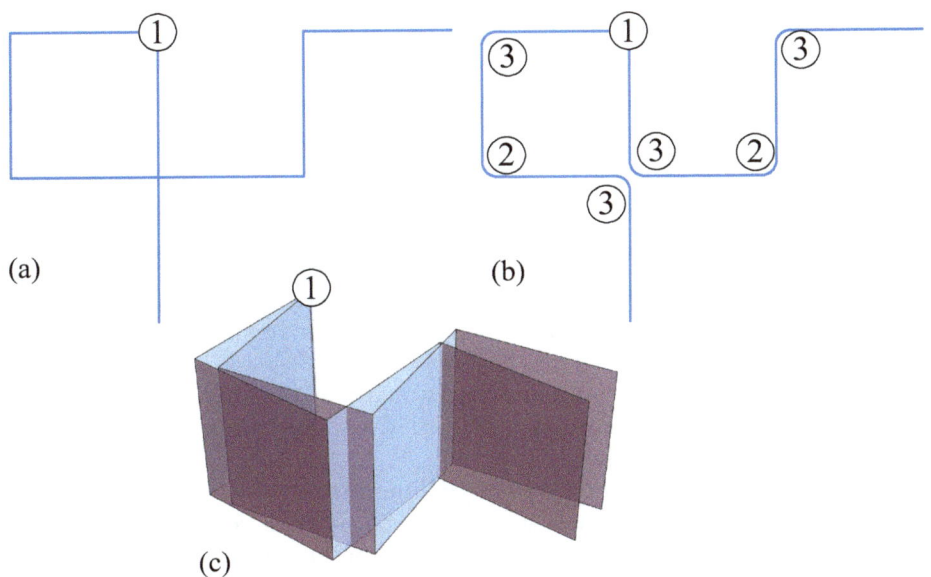

Figure 2.5 (a) Dragon curve of order-3. (b) Rounded corners and crease numbers marked. (c) After nearly closing the first crease.

the first and second creases marked. In (b) and (c) of the figure, the first crease is almost closed. The orthogonal curve in (a) is an order-2 dragon curve; it is order-2 because $n = 2^2 = 4$.

For $n = 8$, the 90° side view becomes an ambiguous curve due to self-touching, as shown in Figure 2.5(a), so usually the curve is displayed with rounded corners as in Figure 2.5(b). We can clearly see that the first halving creases once, the second halving makes two creases, and the third makes four creases. In Figure 2.5(c), crease 1 is nearly closed.

Figure 2.6 Order-10 dragon curve (rounded).

You may wonder why it's called the "dragon" curve. It is because for higher-order curves, with some imagination, it forms a shape reminiscent of a dragon: see Figure 2.6.

The dragon curve has many interesting mathematical properties, of which we'll only mention two: (1) For any n, the order-n dragon curve never self-crosses (but we've seen it self-touches).

(2) The dragon curve is a ***space-filling curve***, in the following sense. First, rescale the curve at each step by $1/\sqrt{2}$. So each edge in the order-10 curve (Figure 2.6) has length $(1/\sqrt{2})^{10} = 1/32$. Second, put four dragon curves together, starting at the same point but each rotated 90° with respect to the previous. See Figure 2.7. Third, in the limit as $n \to \infty$, the four curves fill the infinite plane. This means that if you pick any point in the plane, one of the four infinite, scaled dragon curves covers that point.

2.4 Counting Stamp Foldings

We'll next turn to counting the number of foldings. This is an aspect of combinatorics more than of geometry, but we'll see geometry still plays a role. The natural question is:

> **Q.** How many different ways can a strip of n unit-square stamps be folded to a 1×1 stack?

2.4. Counting Stamp Foldings

Figure 2.7 Four intertwining dragon curves. [Wikimedia Commons, author Stefan Lew.]

There are several notions of what constitutes "different ways." Let us label the stamps $1, 2, \ldots, n$ and count the different orderings of the stamps in the stack, top-to-bottom. So the ordering 132 is considered different from 231. Under this interpretation, the answer to the counting question seems obvious: $n!$, the number of permutations of n numbers (Box 2.3). So for $n = 3$, there are $3! = 6$ permutations:

$$123, 132, 213, 231, 312, 321.$$

Box 2.3 Permutations

Permutations count the number of different orderings of a set of n objects. The number is expressed using the factorial notation $n! = n \cdot (n-1) \cdots 3 \cdot 2 \cdot 1$. So $4! = 24$.

The function $f(n) = n!$ grows very fast. It is larger than 2^n (for $n \geq 4$) and smaller than n^n (for $n \geq 2$). It grows faster than c^n for any constant $c > 1$.

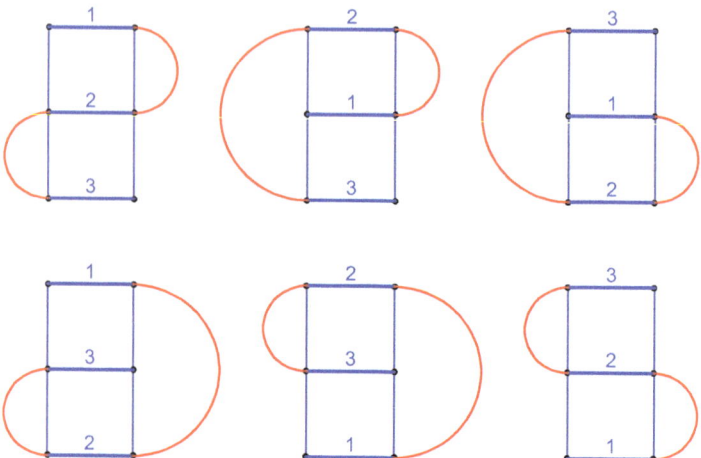

Figure 2.8 $3! = 6$ permutations of 3 stamps. Stamps are the horizontal segments viewed edge-on.

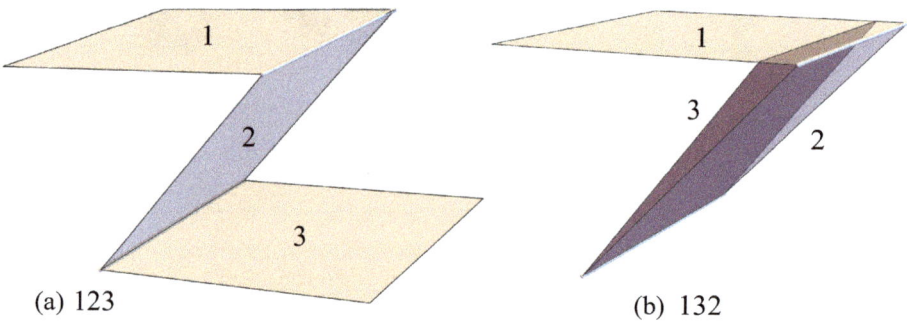

Figure 2.9 Foldings 123 and 132 in Figure 2.8.

> **Exercise 2.2 [Practice] $n!$ Growth**
>
> Argue that $2^n < n! < n^n$, for $n \geq 4$.

> **Exercise 2.3 [Understanding] $n!$**
>
> Argue that the number of permutations of n objects is n times the number of permutations of $n-1$ objects.

We can display the corresponding stamp foldings in a side view as in Figure 2.8. In this depiction, the red arcs represent the perforated connection between adjacent stamps. So the upper left 123 is the pleat M/V fold shown in Figure 2.9. The lower left 132 tucks stamp 3 between 1 and 2, as crudely depicted in Figure 2.9(b).

2.4. Counting Stamp Foldings

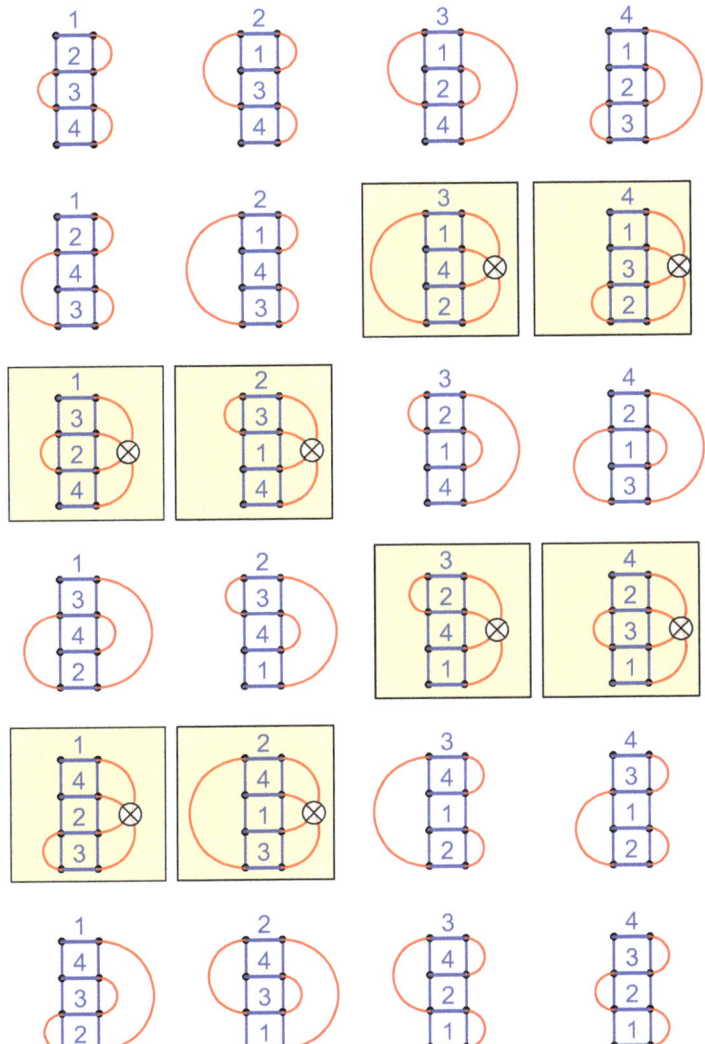

Figure 2.10 4! = 24 permutations of 4 stamps. Those marked with ⊗ are not proper foldings due to paper penetration.

But already for $n = 4$ stamps, the number of different foldings is less than $4! = 4 \cdot 3 \cdot 2 \cdot 1 = 24$, as demonstrated in Figure 2.10. Consider the permutation 1324, the third in column 1, which compared to the pleat folding 1234, swaps the positions of stamps 2 and 3. This requires the connection between stamps 1 and 2 to cross through the connection between 3 and 4, only possible by cutting the strip of stamps into two strips. Eight of the 24 permutations have similar crossings, so there are only 16 different foldings of $n = 4$ stamps.

What is the general pattern for $f(n)$, the number of foldings of a strip of n labeled stamps? Remarkably, this remains unknown, despite considerable efforts by many mathematicians (and amateurs).

> **Open Problem 2.1 Stamp Foldings**
>
> Find an exact closed-form formula for $f(n)$, the number of labeled foldings of a strip of n unit-square stamps to a 1×1 stack.

The number has been calculated to quite large n, but no precise pattern ("closed-form formula") has emerged. The sequence is A000136 in the *Online Encyclopedia of Integer Sequences.*[1] It begins

$$1, 2, 6, 16, 50, 144, 462, 1392, 4536, 14060, \ldots$$

with 6 and 16 corresponding to Figures 2.8 and 2.10 respectively.

We know $f(n)$ is bounded above by $n!$: $f(n) \leq n!$. An easy lower bound follows because each crease can be either a mountain M or a valley V crease. So there are 2 choices for each of the $n-1$ creases, leading to $f(n) \geq 2^{n-1}$. For example, for $n = 3$, this says that there are at least $2^2 = 4$ different foldings; we've seen there are in fact 6: Figure 2.8. Counting by M/V numbers is rather different from counting by permutations, as the M/V pattern ignores the tuck orderings of the stamps and also ignores stamps illegally crossing.

These upper and lower bounds are rather loose. But in fact $f(n)$ is quite close to 3.3^n, as illustrated in the log plot in Figure 2.11. So even though the exact count is not known to follow a formula, the growth of the count is quite regular. In fact, it has been proved that $f(n)$ grows exponentially, somewhere between 3^n and 4^n. We'll record that as a theorem for later reference.

> **Theorem 2.1 Stamp Folding Count**
>
> The number of different ways $f(n)$ a strip of n unit-square labeled stamps can be folded to a 1×1 stack grows exponentially with n: $3^n < f(n) < 4^n$ (for sufficiently large n).

The hedging "(for sufficiently large n)" means that the claimed relationship holds for all n beyond a certain value. In our case n has to reach 25 before the lower bound holds: $3^{25} < f(25)$.

2.5 Map Folding

Stamps are frequently organized into sheets. The goal is the same: Fold into a 1×1 stack of square stamps.

---

[1] https://oeis.org/A000136.

2.5. Map Folding

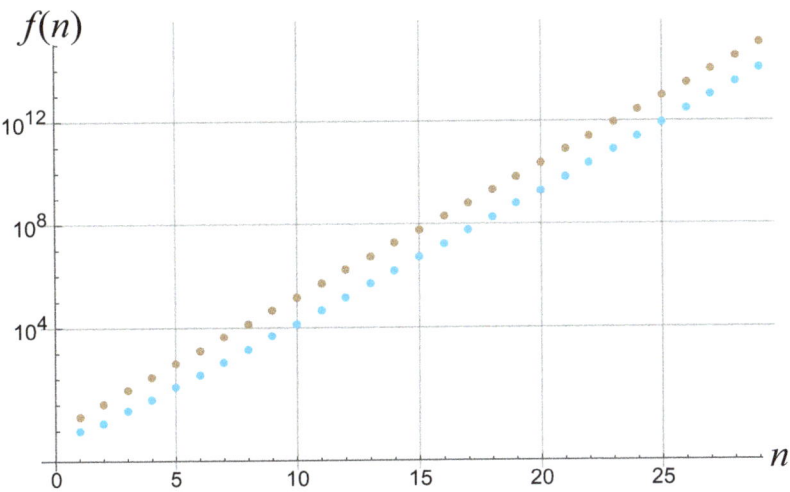

Figure 2.11 Log plot of $f(n)$. Blue dots: $f(n)$, brown dots: 3.3^n.

Figure 2.12 Sheet of postage stamps. [British government, Public domain, via Wikimedia Commons]

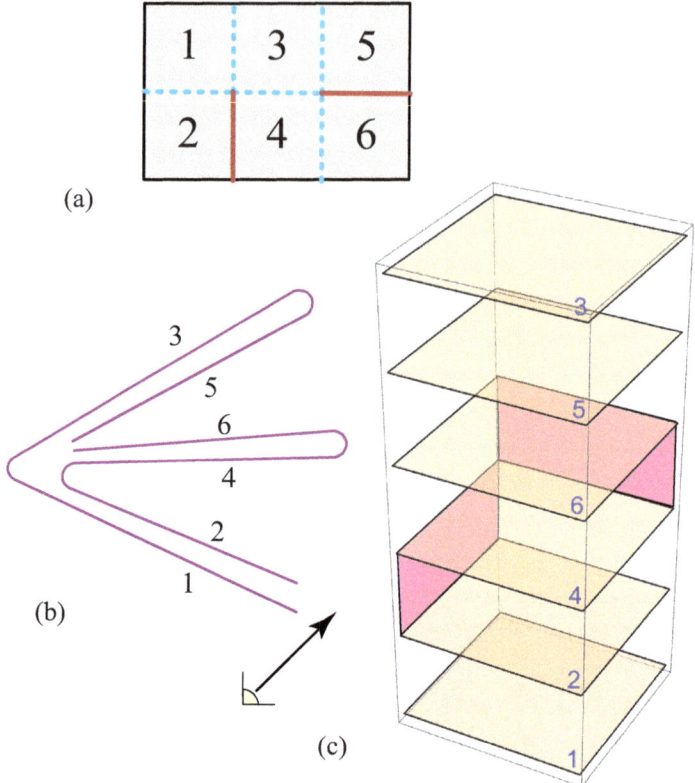

Figure 2.13 Folding a 3×2 map to a stack of squares. (a) The map. (b) Side view of square edges after folding. (c) 3D stack of squares, with the two mountain-fold connections highlighted.

There is an easy strategy, which we'll illustrate with the sheet in Figure 2.12. Fold each of the four horizontal rows as a unit via a pleat folding, reducing the 5 × 4 sheet to a 5 × 1 strip, four stamps thick. Now we've reduced a 2D (two-dimensional) problem to 1D (one-dimensional), and we can then follow strip-folding approaches.

However, of greater interest than counting the number of foldings, is what is known as the ***map folding problem***. Prior to GPS availability, road maps were often essential for traveling by car. After consulting an open map, it could be difficult to refold it to its original compact shape, following the original M/V crease assignment, to restore it in the car's glove compartment.

So this raised the following question:

Q. Given an $n \times m$ map formed of unit squares, with a given M/V assignment for every crease (i.e., for every edge shared by two squares), can it be folded to a 1×1 stack of squares?

2.5. Map Folding

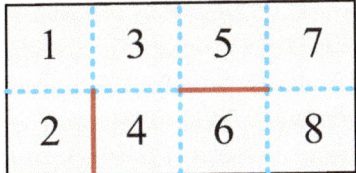

Figure 2.14 A 4×2 map to fold.

The challenge is to avoid trying every possible folding consistent with the M/V assignment to determine the answer, for there are an exponential number of such possibilities. If only a polynomial number of foldings need be explored, then we'd have what is known as a ***polynomial-time*** algorithm (Box 2.4).

To give a sense of the difficulties, even folding the 3×2 map in Figure 2.13 is not straightforward. It can be achieved by first V-folding 5/6 on top of 3/4, resulting in a 2×2 square. Then V-fold this in half along the horizontal midline, producing a 2×1 rectangle. Finally M-fold in half. It's not obvious but this achieves the M/V crease assignments.

> **Exercise 2.4 [Challenge] 4×2 Map**
>
> Cut the 4×2 map in Figure 2.14 out of paper, and fold it into a 1×1 stack, obeying the M/V assignment.

> **Box 2.4 Polynomial Time**
>
> We've earlier seen the distinction between exponential growth and polynomial growth of functions of n in Box 2.1. Algorithm complexity is crudely measured as exponential or polynomial as a function of the input size of the problem, as usual represented by n. An algorithm is said to be ***polynomial time*** if its running time as $n \to \infty$ ("asymptotically") is upper-bounded by some constant times a polynomial in n. We've seen that even if the polynomial is n^{100}, eventually ("asymptotically") it outperforms an exponential-time algorithm.
>
> Later (in Chapter 4) we will introduce the NP-hard problems and how they relate to exponential-time algorithms.

We will revisit this map-folding question in Chapter 5, but for now we note that the problem has only been solved for $n \times 2$ maps, leaving this open:

> **Open Problem 2.2 Map Folding**
>
> For an $n \times m$ map of squares, $m > 2$, and a given M/V assignment to all edges, is there a polynomial-time algorithm to determine whether the map can be folded to a 1×1 stack of squares?

2.6 Technical Notes

Sec. 2.1: Introduction Much of my presentation in this chapter follows Ryuhei Uehara's Chs. 5 and 6 in Uehara (2020).

Sec. 2.2: Exponential Growth For further details, see any undergraduate algorithms textbook.

Sec. 2.3: Dragon Curve Here I follow Uehara (2020, Ch. 6) and Tabachnikov (2014).

Sec. 2.4: Counting Stamp Foldings The plot in Figure 2.11 is from Uehara (2020, Figure 5.6). Chapter 6 of Uehara (2020) analyzes the computational complexity of several stamp-folding problems.

Sec. 2.5: Map Folding For a more difficult but still foldable 3×3 map, see Demaine and O'Rourke (2007, Figure 14.1). The thesis that established that $n \times 2$ map folding is polynomial time is Morgan (2012).

3

Flat Vertex Folds

3.1 Introduction

Although typically the goal of an origami construction is a 3D object—perhaps a box, or a jumping frog, or the iconic crane—intermediate stages often have flat components. In this chapter we study "single vertex flat folds," one of the few corners of the origami world whose mathematics is thoroughly understood. A ***vertex*** is a point v surrounded by 360° of origami paper, to which are incident creases, segments with one endpoint at v. The number of incident creases is called the ***degree*** of v. A single vertex is just that: v is alone in the middle of a piece of paper, whose boundary shape is of no concern. Every origami construction contains usually many vertices, and frequently they are flat or nearly flat. (For example, see ahead to Figure 6.1.) Even if a vertex is not flat, it often follows the conditions for flat-foldability. We'll see in Chapter 5 the importance of flat folding in rigid origami dynamics (Section 5.4.2). Chapter 4 will explore flat folding of crease patterns with many vertices.

Each crease segment incident to the single vertex v is assigned to be an M-fold or a V-fold. The central question is under what conditions can the crease pattern fold flat, obeying the M/V assignment and only bending along the given creases, folding into multiple layers with no tearing of the paper. Repeating terminology, we use ***crease pattern*** for the pattern of creases prior to an M/V assignment, and ***M/V pattern*** after assignment.

A simple example is the result of folding a sheet of paper in half twice: Once top-to-bottom, and then left-to-right, which produces a degree-4 vertex; see Figure 3.1. A more complicated degree-8 example is shown in Figure 3.2.

Not every M/V pattern is flat-foldable, nor does every crease pattern have some M/V assignment that renders it flat-foldable. See Exercise 3.1. So there must be conditions imposed for flat-foldability.

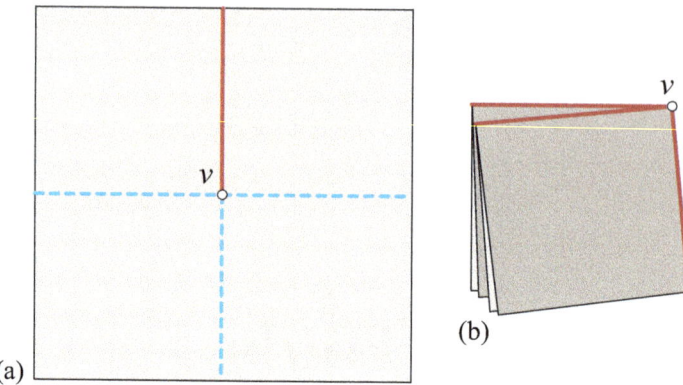

Figure 3.1 Degree-4 vertex: (a) M/V (red/blue) creases on lighter side of paper; backside is darker. (b) Flat folding. The three valley creases become mountain creases on the darker side. [From O'Rourke (2011). Reprinted by permission of Cambridge University Press.]

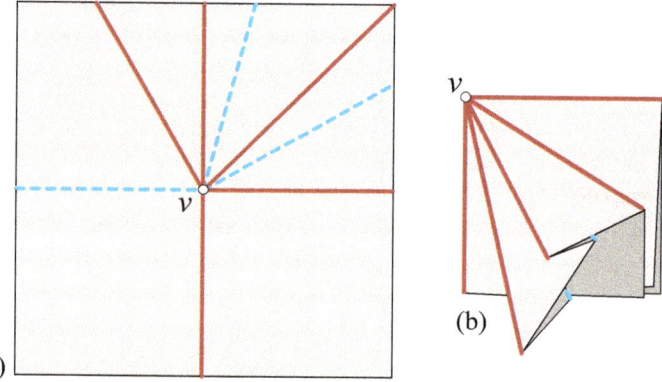

Figure 3.2 Degree-8 vertex: (a) crease pattern. (b) Folding. [From O'Rourke (2011). Reprinted by permission of Cambridge University Press.]

> **Exercise 3.1 [Practice] Not Flat-Foldable**
>
> (a) Create four mountain creases meeting orthogonally at a central vertex v, as shown in Figure 3.3(a). Convince yourself by manipulation that the paper cannot fold flat with just those four creases mountain-folded and meeting at the central vertex v (in contrast to Figure 3.1).
>
> (b) Draw the degree-3 vertex crease pattern shown in Figure 3.3(b), and convince yourself that no M/V assignment to the three creases allows this pattern to fold flat.
>
> (c) Convince yourself that the M/V assignments to the four creases in this example prevent flat folding.

3.2. Maekawa's Theorem

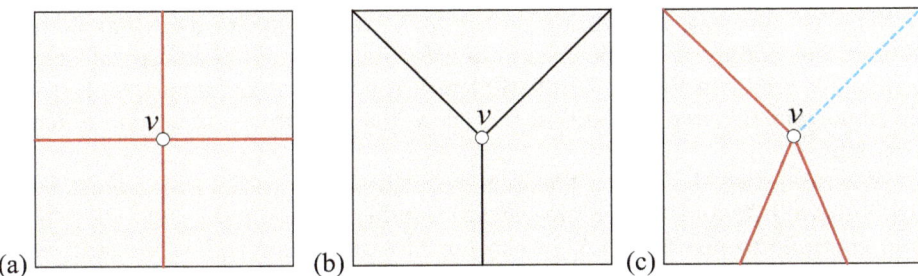

Figure 3.3 Exercise 3.1. (a,c) Four creases with M/V assignments meeting at degree-4 vertex v. (b) Degree-3 crease pattern without M/V assignments.

Conditions come in two varieties. **Necessary** conditions are needed for the possibility of flat-folding. Any pattern that folds flat must satisfy these conditions. As a reader who studied Figure 3.3(b) of Exercise 3.1 might guess, one necessary condition is that the degree of the vertex must be even. **Sufficient** conditions are enough by themselves to ensure flat-foldability. Having an even number of creases incident to the vertex is *not* sufficient, as Figure 3.3(c) shows.

The holy grail of mathematics is complete "characterization" provided by necessary and sufficient conditions. (See ahead to Box 3.2.) This is what is achieved by Kawasaki's Theorem (Section 3.5). But we'll see that Kawasaki's conditions only address crease patterns, not M/V assignments to those patterns. So we first examine three necessary conditions constraining flat-foldable M/V patterns: Maekawa's Theorem 3.1, the Even-Degree Lemma 3.1, and the Local-Min Lemma 3.2.

3.2 Maekawa's Theorem

When attempting to flatten the four M creases in Figure 3.3(a), the paper seems to want one of the M creases to instead be a V crease. This is why the example in Figure 3.1 does fold flat, this time with three V and one M crease, which flipped over becomes three M and one V crease. Exploring further examples, such as the pattern in Figure 3.2 (five M and three V creases), could lead one to conjecture the regularity captured by Jun Maekawa, now known as Maekawa's Theorem:[1]

> **Theorem 3.1 Maekawa**
>
> If M mountain creases and V valley creases meet at a vertex of a flat folding, then M and V differ by 2: either $M = V + 2$ or $V = M + 2$.

---

[1] Also known as the Maekawa–Justin Theorem, as Jacques Justin discovered it independently.

This is a necessary condition on the M/V assignment to allow flat folding. It does not supply a sufficient condition for flat folding: The three M creases and one V crease in Figure 3.3(c) still leaves the pattern unflattenable. But it is an important necessary condition that we will see employed in several different contexts.

We now prove Maekawa's Theorem, referring to Figure 3.4 throughout. We'll use italicized M and V to represent the number of mountain and valley folds, and continue to use M and V to shorten Mountain and Valley. We start with a circular piece of paper (Figure 3.4(a)) so we are not distracted by the paper corners, which are irrelevant to what happens in the **neighborhood** of the single central vertex. Assume it does fold flat. Our goal is to prove that M and V differ by 2.

Lay the folding flat, as in Figure 3.4(b). Now look toward v along the plane into which the folding has been flattened. This view shows a closed zig-zag path of circular arcs, as depicted in Figure 3.4(c). Each arc is a piece of the circular boundary flattened between two creases, which, viewed edge-on, appears as

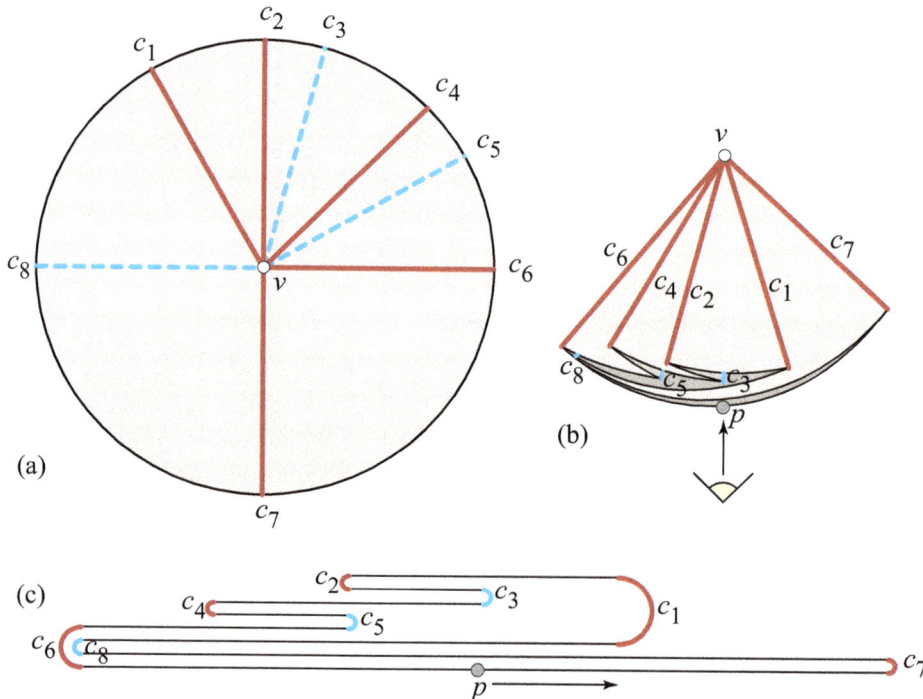

Figure 3.4 The example of Figure 3.2 revisited: (a) M/V crease pattern on circular paper. The eight creases are labeled $c_1, \ldots, c_8$. (b) Flat folding. (c) Expanded view looking from folded boundary toward vertex. Sharp turns at creases are shown as circular arcs to illustrate the nesting. [From O'Rourke (2011). Reprinted by permission of Cambridge University Press.]

3.2. Maekawa's Theorem

a straight segment. Select any point of the path not directly at a crease, for example, point p in Figure 3.4(b,c), and imagine walking toward the right. Let's view your direction of travel as a vector, an arrow pointing in the direction of travel. (Recall Box 1.2.) Then the start direction vector points at angle 0° in the standard coordinate system, in which angles are measured counterclockwise from the positive x-axis, which points toward the right.

Each mountain fold you encounter in your walk rotates your direction heading through $+180°$ ($+$ meaning counterclockwise), and each valley fold rotates your vector through $-180°$ ($-$ for clockwise).

Now, we know that by the time the walk returns to the starting point p after traversing the entire diagram in Figure 3.4(c), we approach p from the left heading right, so again the vector has direction 0°, which is the same as 360°. In other words, we must twist a total of a full 360° by the time we return to start.

So we must have

$$M \cdot 180° + V \cdot (-180°) = 360° .$$

Dividing through by 180° leads to $M - V = 2$. Flipping the paper over interchanges the roles of M and V, and we reach the conclusion that $V - M = 2$. Combining both possibilities into one phrase: M and V differ by 2: $|M - V| = 2$. And that is the exactly the claim of the theorem; so we have proved Theorem 3.1.

Most theorems have many proofs, often starting from different background assumptions. An alternate proof of Maekawa's Theorem using polygons is presented in Box 3.1.

Box 3.1 Proof of Maekawa's Theorem via Polygons

The following proof was found by Jan Siwanowicz when he was a high-school student. The starting point of his proof is another theorem: The sum of the internal angles at the n vertices of a polygon is $(n-2)180°$. The idea is to view the zig-zag path in Figure 3.4(c) as a squashed polygon, as in Figure 3.5, which is closer to how it looks with sharp creases. If we imagine compressing this polygon completely flat, all the mountain vertices have an internal angle near 0°, and all the valley vertices have an internal angle near 360°. So the total internal angle sum after complete flattening is

$$M \cdot 0° + V \cdot 360° ,$$

and this must equal $(n-2)180°$, where n is the total number of vertices of the polygon. In this construction, each vertex derives from a crease, so $n = M + V$. Therefore,

$$V \cdot 360° = (M + V)180° ,$$

and dividing by 180° yields $2V = M + V$ or $M - V = 2$.

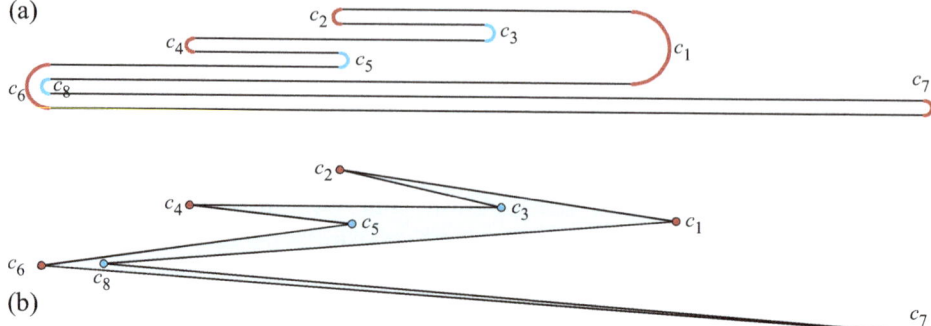

Figure 3.5 An 8-vertex polygon (b) corresponding to Figure 3.4(c), repeated as (a) here, with 5 mountain vertices $\{c_1,c_2,c_4,c_6,c_7\}$ and 3 valley vertices $\{c_3,c_5,c_8\}$. [From O'Rourke (2011). Reprinted by permission of Cambridge University Press.]

> **Exercise 3.2 [Practice] Maekawa's Theorem**
>
> Add additional creases to Figure 3.3(a), retaining the four M creases, so that it can flatten, and verify Maekawa's Theorem 3.1 for your construction.

3.3 Even-Degree Lemma

We already noted (without proof) that it is necessary for a single vertex crease pattern to fold flat, that the vertex have even degree, i.e., its **parity** is even. We can use Maekawa's Theorem to prove this:

> **Lemma 3.1 Even-Degree**
>
> A vertex in a flat folding has even degree.

If a crease pattern is flat-foldable, then an M/V assignment that achieves that flat folding must satisfy $M = V + 2$. Then

$$M + V = (V+2) + V = 2V + 2 = 2(V+1)$$

and so the total number of creases $M+V$ incident to vertex v is even: 4 in Figure 3.1, 8 in Figures 3.2 and 3.4. Starting with $V = M + 2$ reaches the same conclusion: $M+V$ is even.

Sometimes the Even-Degree Lemma is phrased as claiming that the sectors (regions between consecutive creases) surrounding a flat-foldable vertex v can be **2-colored**: meaning that the sectors can be colored so that no two adjacent sectors receive the same color.

3.4 Local-Min Lemma

The pattern in Figure 3.6(a) satisfies the necessary conditions in the previous two sections: v has degree-6 and so satisfies the Even-Degree Lemma 3.1, and $M = 4$ and $V = 2$ satisfies Maekawa's Theorem 3.1.

And yet, if you try to fold it flat, you will see it is impossible. Why? The essence of the impediment is that a 40° pie-slice sector delimited by two valley folds is surrounded by larger angles on either side—70°. This forces paper to pass through itself, as depicted in Figure 3.6(b). Whenever we have such a pattern of consecutive sector angles—large, small, large—the folds delimiting the central sector cannot both be valley, nor both mountain: One must be M and the other V. The central angle is called a ***local-min***, because locally, that is, in its immediate neighborhood, it is a minimum angle, smaller than its neighbors to either side. To be more precise, a ***strict local-min*** angle is strictly less than ($<$) on both sides. We can phrase this condition in a lemma as follows:

> **Lemma 3.2 Local-Min**
>
> In any flat folding, any sector whose angle is a strict local-min must be delimited by one mountain and one valley fold.

This is sometimes called the "Big-Little-Big" lemma. We will see it can be derived from Kawasaki's Theorem 3.2.

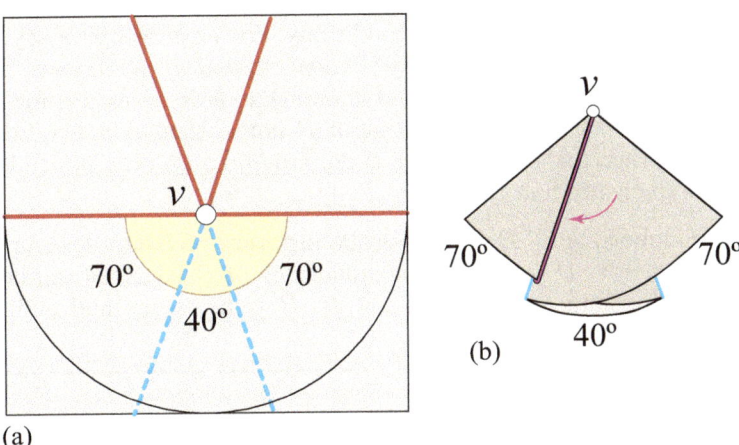

Figure 3.6 (a) An M/V pattern that cannot fold flat. (b) Attempting to fold the 40° sector results in paper penetrating itself. [From O'Rourke (2011). Reprinted by permission of Cambridge University Press.]

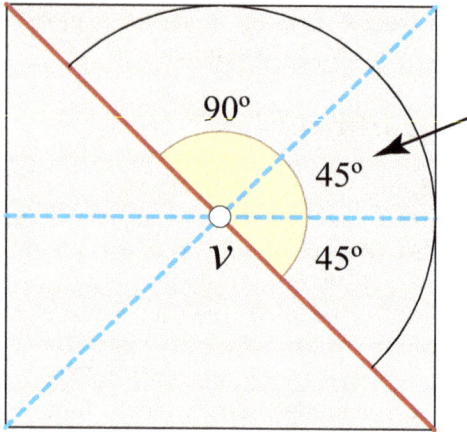

Figure 3.7 Can this M/V pattern fold flat?

> **Exercise 3.3 [Understanding] Local-Min Ties**
>
> Figure 3.7 shows an M/V pattern where a 45° angle lies between 45° on one side and 90° on the other side, delimited by two V creases. Can the pattern fold flat?

> **Box 3.2 If and Only If**
>
> The phrase "if and only if" (often abbreviated "iff") is a shorthand to express necessary and sufficient conditions. For example, let's say A is "folds flat" and B is "even parity." Then "A *only if* B" means "folds flat only if even." This is true, and shows that B is *necessary* for A to hold. Whereas "A *if* B" means that if B holds, then A holds, i.e., "if even parity, then folds flat," which is not true because even parity is not *sufficient* to ensure flat folding.
>
> In logic symbols, $A \Rightarrow B$ expresses the necessity of B for A: A implies B. Whereas $A \Leftarrow B$ expresses the sufficiency of B for A: B implies A. Together, A if and only if B: $A \Leftrightarrow B$, B is necessary and sufficient for A.

3.5 Kawasaki's Theorem

We come now to the promised characterization of flat-foldable crease patterns. Call the sector angles around the vertex in sequential order, $\theta_1, \theta_2, \ldots, \theta_n$. We

3.5. Kawasaki's Theorem

know from the Even-Degree Lemma 3.1 that n is even, because an even number of creases determine an even number of sectors. We also know that

$$\theta_1 + \theta_2 + \cdots + \theta_n = 360°$$

because the angles completely surround the vertex. Kawasaki's Theorem[2] claims that a simple condition on the angles, completely ignoring M/V patterns, provides necessary and sufficient conditions for flat-foldability:

> **Theorem 3.2 Kawasaki**
>
> An even number of creases meeting at a vertex folds flat if, and only if, the alternating sum of the determined sector angles is zero:
>
> $$\theta_1 - \theta_2 + \theta_3 - \theta_4 + \cdots + \theta_{n-1} - \theta_n = 0°.$$

The term **alternating sum** means that every other term has opposite sign: The odd terms $\theta_1, \theta_3, \theta_5, \ldots$ are added and the even terms $\theta_2, \theta_4, \theta_6, \ldots$ are subtracted. So the alternating-sum equation is equivalent to

$$\theta_1 + \theta_3 + \theta_5 + \cdots = \theta_2 + \theta_4 + \theta_6 + \cdots :$$

The sum of the odd-indexed angles equals the sum of the even-indexed angles. (And so both sum to 180°.) Note the odds-equals-evens sums are independent of the starting index: Any sector angle can be chosen as θ_1. See Box 3.2 for the meaning of "if and only if."

> **Exercise 3.4 [Practice] Local-Min and Kawasaki's Theorem**
>
> (a) Verify that the sector angles in Figure 3.6(a),
>
> $$(70°, 40°, 70°, 70°, 40°, 70°),$$
>
> satisfy Kawasaki's Theorem.
>
> (b) Then why doesn't Kawasaki's Theorem contradict the Local-Min Lemma?

The claim that Kawasaki's Theorem 3.2 provides a complete characterization of flat-foldability is rather remarkable, because it says nothing explicitly about the pattern of mountain and valley folds, or the layering order. But because its conditions are sufficient, the alternating angle sum implies both Maekawa's Theorem 3.1 and the Local-Min Lemma 3.2, as we will see shortly.

We illustrate the theorem with the degree-6 example in Figure 3.8(a) with six sector angles

$$31° + 39° + 41° + 58° + 108° + 83° = 360°.$$

---

[2] Also known as the Kawasaki–Justin theorem, as Justin discovered it independently.

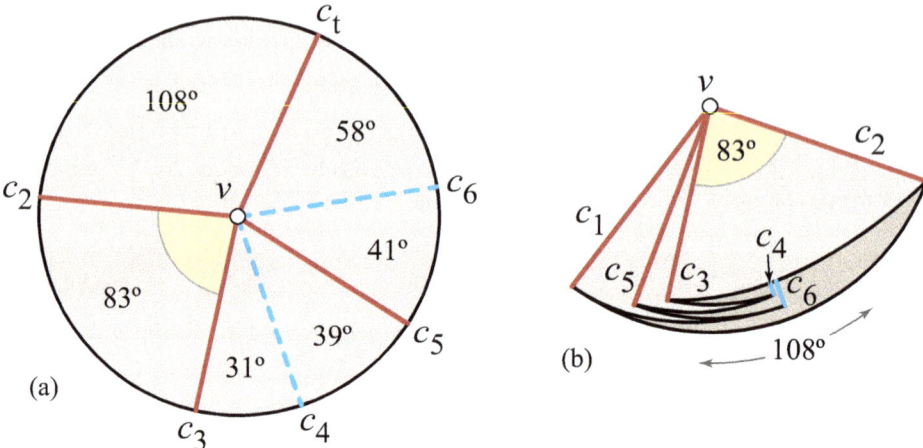

Figure 3.8 Illustration of Kawasaki's Theorem 3.2. [From O'Rourke (2011). Reprinted by permission of Cambridge University Press.]

Their alternating sum is indeed zero:

$$31° + 41° + 108° = 180° = 39° + 58° + 83°,$$

so

$$31° - 39° + 41° - 58° + 108° - 83° = 0°.$$

The flat folding guaranteed to exist by the theorem is shown in Figure 3.8(b).

3.5.1 Kawasaki Necessity

We now prove that if the crease pattern is flat-foldable, then the alternating sum is zero, i.e., that a zero alternating sum is a necessary condition for flat-foldability. The proof proceeds just as with the Maekawa argument, analyzing the zig-zag circular paper boundary path, as in Figure 3.4(c). Again we imagine walking around this path. But now rather than concern ourselves with the gyrations of the direction vector of travel, we concentrate on how far we travel, measuring "how far" not in terms of linear distance, but in terms of angular travel as seen from the central vertex.

Let's use Figure 3.8(b) as an example. Starting at the leftmost edge c_1 of the folding and traveling rightward on the bottommost flap, we travel an arc of 108° with respect to the vertex v. At the mountain fold c_2 we reverse direction and travel an arc of 83° leftward, then reverse again at c_4 and travel 31° rightward, and so on. Whether we encounter a mountain or a valley fold is irrelevant, as is the tucking/layering order: We are only concerned with total angular travel. By the time we return to the start point, the total travel must be 0°. And so the alternating sum is necessarily zero.

3.5. Kawasaki's Theorem

3.5.2 Kawasaki Sufficiency

Now we turn to sufficiency: If the alternating sum is zero, then there is an M/V assignment to the creases that leads to flat folding. This is a somewhat difficult proof, so we proceed slowly, starting with the simple example in Figure 3.9, and later analyze a larger example in Figure 3.11.

The proof employs three key ideas:

(1) Cut one crease incident to the vertex v so that the circular order around v becomes a linear order, beginning and ending at the two sides of the cut. In Figure 3.9(a), the cut labeled 1 results in the counterclockwise angle sequence

$$45° - 90° + 135° - 90° = 0° .$$

(2) Now **accordion fold** (or accordion "pleat") at the creases, alternating M and V, stacking the sector angle lengths vertically. This is represented

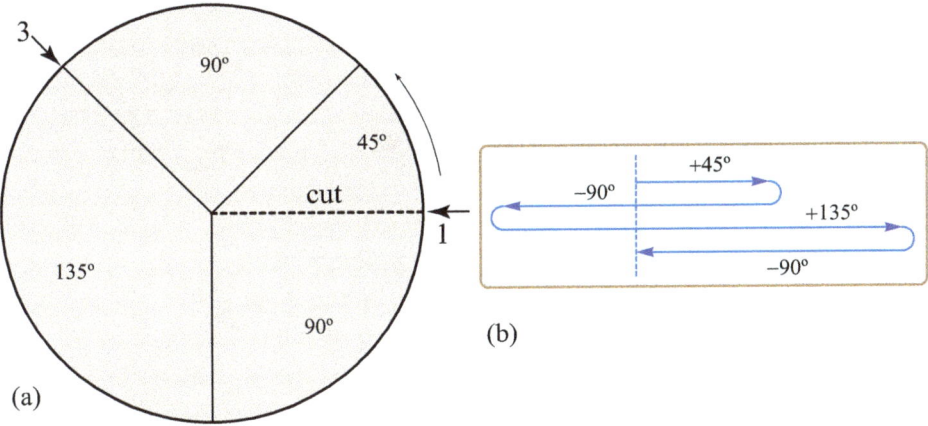

Figure 3.9 (a) Cut at dashed crease. (b) Accordion fold. Dashed vertical shows alternating sum is zero.

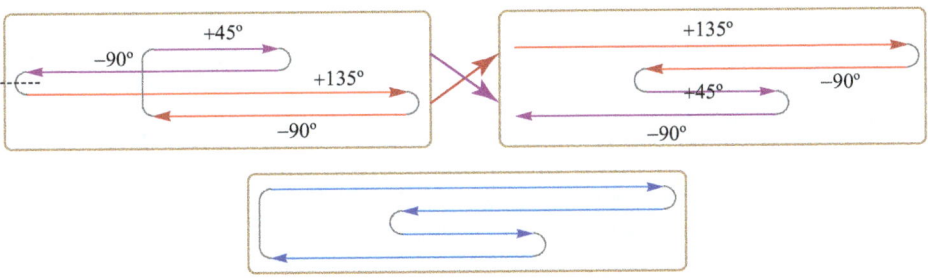

Figure 3.10 Rewiring after cutting at the leftmost excursion (dashed) permits closing without crossing.

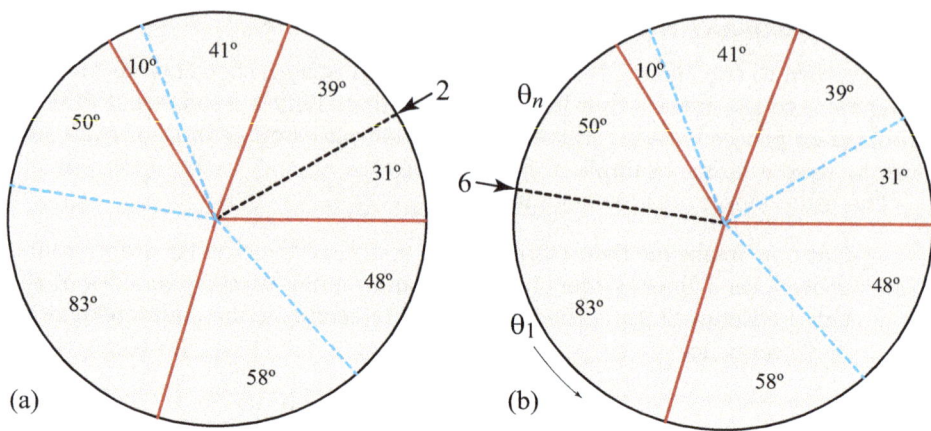

Figure 3.11 (a) Cut at crease 2. (b) Cut at crease 6. The angle θ_1 is counterclockwise of the cut, and labels proceed counterclockwise to θ_n just prior to the cut.

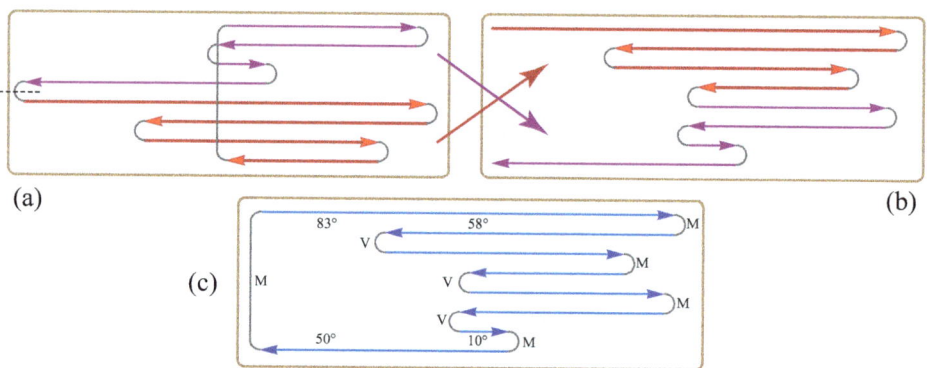

Figure 3.12 (a) Accordion fold cannot close without self-intersection. Cut dashed. (b) Reordering the sequence based on the leftmost excursion. (c) The final M/V assignment.

in Figure 3.9(b). That the alternating sum is zero means that the start and end points align vertically.

(3) Rearrange the cut if necessary so that the accordion fold can close without paper collision. Here we change the cut to crease 3, which results in the reordered sequence shown in Figure 3.10:

$$135° - 90° + 45° - 90° = 0° \ .$$

It is this third step that is tricky. We illustrate it with a more complex example in Figures 3.11 and 3.12.

3.5. Kawasaki's Theorem

The alternating sum of the sector angles in Figure 3.11 is zero:
$$39° - 41° + 10° - 50° + 83° - 58° + 48° - 31° = 0°.$$

If we cut at index 2 (where index 1 is horizontal right of v in Figure 3.11(a)), the accordion fold cannot close: Figure 3.12(a). So now partition the accordion-fold stack at the leftmost excursion, and cut between the leftward horizontal arrow ($-50°$, purple) and the rightward arrow ($+83°$, red). Now swap the purple above and the red below, reordering the sequence to
$$83° - 58° + 48° - 31° + 39° - 41° + 10° - 50° = 0°.$$

This reordering guarantees that the partial alternating sums are all nonnegative:
$$83° - 58° = 25°$$
$$83° - 58° + 48° = 73°$$
$$83° - 58° + 48° - 31° = 42°$$
$$83° - 58° + 48° - 31° + 39° = 81°$$
$$83° - 58° + 48° - 31° + 39° - 41° = 40°$$
$$83° - 58° + 48° - 31° + 39° - 41° + 10° = 50°$$
$$83° - 58° + 48° - 31° + 39° - 41° + 10° - 50° = 0°.$$

This means that the back-and-forth excursions never go left of the starting point—the leftmost excursion—and so it is possible to close the accordion fold on the left without intersection: Figure 3.12(c). This completes the sufficiency proof of Kawasaki's Theorem.

Because Kawaski's Theorem is a complete characterization based only of the crease pattern angles, it must imply the three necessary conditions we discussed. Next we describe the three implications.

Even-Degree Lemma 3.1 This is part of the premise of Kawasaki's Theorem.

Local-Min Lemma 3.2 Because the sufficiency proof uses an accordion fold, the M/V assignment alternates M and V. So all the angles but two—the first θ_1 and the last θ_n—are delimited by one M and one V, satisfying the Local-Min Lemma. In the example, $\theta_1=83°$ and $\theta_n=50°$. The final M/V assignment results in three consecutive M-folds surrounding θ_1 and θ_n, three because the cut itself becomes a mountain fold to reclose the sequence. But as is evident in Figure 3.12(c), it must be that $\theta_1 \geq \theta_2$ ($83° > 58°$) and $\theta_n \geq \theta_{n-1}$ ($50° > 10°$), as otherwise the cut was not at the leftmost excursion. So neither θ_1 nor θ_n violates the Local-Min Lemma.

Maekawa's Theorem 3.1 As just mentioned, the M/V assignment alternates M and V, except for the three consecutive M-folds closing the cut. So $M = V + 2$.

Although the sufficiency proof of Kawasaki's Theorem provides a specific M/V assignment, there are many other assignments that also fold flat. It may come as no surprise that there are in fact exponentially many different foldings:

The pleat folding of a strip of stamps (Figure 2.2) is analogous to the accordion folding provided by Kawasaki's Theorem. The specific exponential growth is $\geq 2^{n/2}$ for a vertex of degree n, a result we claim but do not further justify.

> **Exercise 3.5 [Understanding] Exponentially Many M/V**
>
> The vertex in Figure 3.9(a) is surrounded by $n=4$ sector angles, so there should be (at least) $2^{n/2} = 4$ M/V assignments. Find four M/V assignments that each fold flat.

Kawasaki's Theorem can stand as a model of the type of mathematical understanding we seek: a simple but complete characterization of a certain origami behavior. It is not often or easily achieved, but we will see several examples, including the Degree-4 Folding Theorem 5.2, the Fold & 1-Cut Theorem 7.1, and the curved-crease Osculating Plane Bisection Theorem 8.1 (among others).

In the next chapter we move from flat folding single vertices to flat folding multiple vertices, and we'll see a considerable jump in complexity.

3.6 Technical Notes

Sec. 3.1: Introduction Much of this chapter relies on Demaine and O'Rourke (2007, Ch. 12) and O'Rourke (2011, Ch. 4). The topic is also well covered by Lang (2017, Ch. 1) and Hull (2020, Ch. 5).

Sec. 3.2: Maekawa's Theorem See Demaine and O'Rourke (2007, Sec. 10.2) for a concise history of origami mathematics. Both Lang (2017) and Hull (2020) include extensive historical remarks throughout their books.

Sec. 3.5: Kawasaki's Theorem The proof details are drawn from Demaine and O'Rourke (2007, Sec. 12.2.2). The $\geq 2^{n/2}$ result is due to Hull (2020, Thm. 5.26). Hull also poses two single-vertex open problems, so even this "elementary" topic is not entirely settled.

4

Flat Folding Is Hard

4.1 Introduction

We saw in the previous chapter that conditions for flat folding a single-vertex crease pattern are well understood mathematically: Maekawas Theorem 3.1, the Even-Degree Lemma 3.1, the Local-Min Lemma 3.2, and Kawasaki's Theorem 3.2. However, flat-foldability of crease patterns with multiple vertices is not as well understood. And in fact there is strong evidence that there is no characterization analogous to the clean, simple lemmas and theorems determining single-vertex flat-foldability. How can such a negative result be established? Perhaps surprisingly, through computer science.

The past several decades have seen an exciting infusion from computer science theory into what is now called ***computational origami***, investigating the computational complexity of origami constructions and questions. The term ***computational complexity*** refers to how long it takes for an algorithm to solve a problem or resolve a question—if a very long time, the implication is that the problem is computationally complex and would resist a simple characterization. In this chapter we discuss two results in computational origami: flat-foldability is hard—technically, "NP-hard" (Section 4.3), and flat-foldability is richly complex—technically, "Turing-complete" (Section 4.5). Both results are intricate and beyond what we can present formally. But the intuition behind both proofs can be grasped. This will require two detours into theoretical computer science before detailing some origami "gadgets": small crease patterns whose flat-folding properties play specific roles.

This chapter will introduce three triangle gadgets. The first cannot fold flat (Section 4.3.1), so it is not really a "gadget" because it doesn't do anything, but it is an instructive example. The second triangle gadget serves to duplicate a "truth value" in the NP-hard proof (Section 4.3.5). And the third triangle gadget twists, negates, and duplicates a different type of truth value in the Turing-completeness proof (Section 4.5.2). This last gadget will prepare us for the square twist and related foldings in Chapter 5 (Section 5.4.4).

4.2 P, NP-Complete, NP-Hard

To elucidate the main idea behind the concept "NP-hard," we return to strip folding (Chapter 2), posing an admittedly contrived question to illustrate what it means for a problem to be NP-hard. We defer defining the term NP-hard to Section 4.2.2.

4.2.1 Set Partition Is NP-Hard

Suppose we have a $m \times 1$ strip of paper, marked with lines indicating possible creases. The distances between the lines are positive integers; call the set of these numbers S. In Figure 4.1(a), the distances between creases are the eight numbers

$$S = 4,2,7,5,1,8,3,6\ .$$

The task is to fold the strip so that it becomes exactly half as long. The folding may only use the given possible creases, each of which can be folded M or V or left uncreased. In the literature, such creases are called *optional* because they might remain uncreased. The strip in Figure 4.1(a) has total length 36, and indeed it can be folded to length 18 as illustrated in Figure 4.1(b):

$$4+8+6 = 18 = 2+7+5+1+3\ .$$

This folding partitions set S into two equal-length halves $S_1 = \{4,8,6\}$ and $S_2 = \{2,7,5,1,3\}$. Note in Figure 4.1(b) that the elements of S_1 "aim" rightward, and those in S_2 aim leftward.

Abstracting away from strips and creases, the question is known as the SET PARTITION problem:

> Given a set S of n positive integers whose sum is m, can S be partitioned into two sets S_1 and S_2 such that the sum in each of these sets is $m/2$?

Although it may seem in some sense straightforward to solve an instance of this problem, in fact no one knows how to answer this question quickly, and it gets harder and harder as n increases. Even for the seven numbers in Figure 4.1(c), it is not easy to see that the answer is NO—I simply tried all 128 partition possibilities and none partitioned into $18 + 18 = 36 = m$.

Exercise 4.1 [Understanding] Set Partition

Find set partitions for:

(a) $S = \{1,2,3\}$

(b) $S = \{1,2,3,4\}$

(c) $S = \{1,2,3,5,6,7\}$.

(Neither $\{1,2,3,4,5\}$ nor $\{1,2,3,4,5,6\}$ have set partitions.)

4.2. P, NP-Complete, NP-Hard

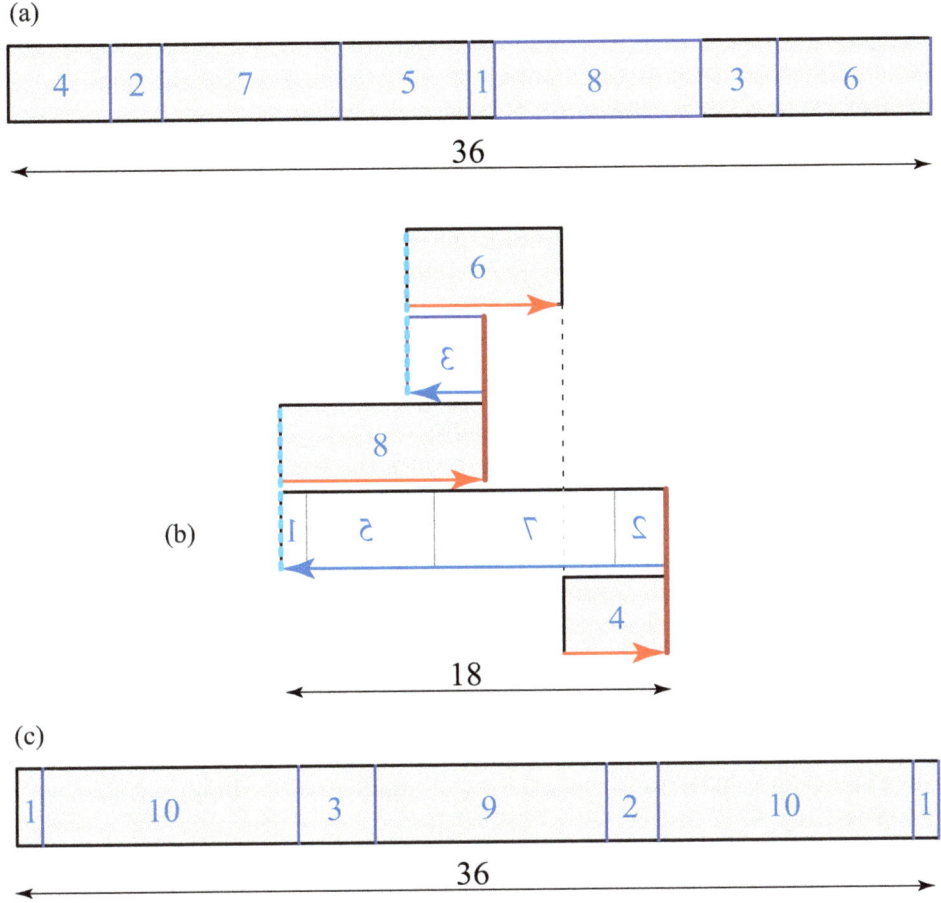

Figure 4.1 (a) A strip that can fold into half its length, $18 = 36/2$. Back side is white. (b) Note after the first M crease between 4 and 2, three possible crease lines are left uncreased. (c) A strip that cannot be partitioned into two equal halves.

Let's examine the simple try-all-possibilities algorithm for answering an instance of SET PARTITION. Write out the n elements of S in a row, with a 1 or 0 above each number indicating whether that number is in S_1 or in S_2 respectively. So, the solution for the Figure 4.1(b) example is as follows.

1	0	0	0	0	1	0	1
4	2	7	5	1	8	3	6

One try for the Figure 4.1(c) example leads to $S_1 = 16$ and $S_2 = 20$.

0	0	1	0	1	1	1
1	10	3	9	2	10	1

Every possible partition of S is then determined by such a 0/1 *bit vector*. As there are 2^n bit vectors of length n—two choices per element—this algorithm requires 2^n steps to try all possibilities. Thus, recalling Box 2.1, this algorithm's number of steps grows *exponentially* with respect to n.

As mentioned above, for Figure 4.1(c), the algorithm checks $2^7 = 128$ partitions. Checking 2^7 partitions is easy, but if S contains, say, $n = 100$ numbers, then with $2^{100} > 10^{30}$, a brute-force search for a set partition is quite infeasible even on the fastest computers.

4.2.2 Complexity Hierarchy

The area of computational complexity is relatively recent, with key aspects initiated in the 1970s. By now the field has blossomed to identify many different classes of complexity, so many (over 500) and so varied that they have been collectively called the "Complexity Zoo." With the reader's indulgence, we will only distinguish two levels of complexity: easy or hard, tractable or intractable. The technical names for these classes are P and NP-hard respectively. Problems in the class P can be solved in *polynomial time*, which means that the needed run time grows no faster than some polynomial in the input size n. As we discussed in Section 2.2, exponential growth always eventually outruns polynomial growth, so there is justification for viewing polynomial growth as "easy." Polynomial-time run time is upper-bounded by cn^k for some constant c dependent on the speed of the computer.

Here is an example, called 3-SUM. Given a set of n integers,[1] determine if any three sum to 0. Here is the same set of numbers used above, but now with 7 replaced by -7.

$$S = 4, 2, -7, 5, 1, 8, 3, 6 \ .$$

Checking all triples of numbers drawn from S finds several solutions, e.g., $4 - 7 + 3 = 0$. If instead 2 is replaced by -2, no triple of numbers from S sum to 0. It should be clear that the growth rate of the check-all-triples algorithm is cn^3—cubic. (With some care one can solve the problem with a quadratic algorithm.) But even if we could only establish an upper bound of n^{100}, it would still be considered easy or tractable.[2]

In contrast, the class of problems that are NP-hard are considered intractable: They are at least as hard as (and maybe harder than) the NP-complete problems. These are a collection of problems proven to be the hardest in the class NP. Although one might think that "NP" stands for "nonpolynomial," in fact it abbreviates *Nondeterministically Polynomial*. It would take us far afield to explain this adequately, so we'll just rest content with the informal distinction: P = easy/tractable vs. NP-complete and NP-hard = hard/intractable. See Figure 4.2 for the relationships among the

---

[1] Recall from Box 1.1 that integers include negative numbers.
[2] One of my algorithms has an upper bound of n^{42}.

4.2. P, NP-Complete, NP-Hard

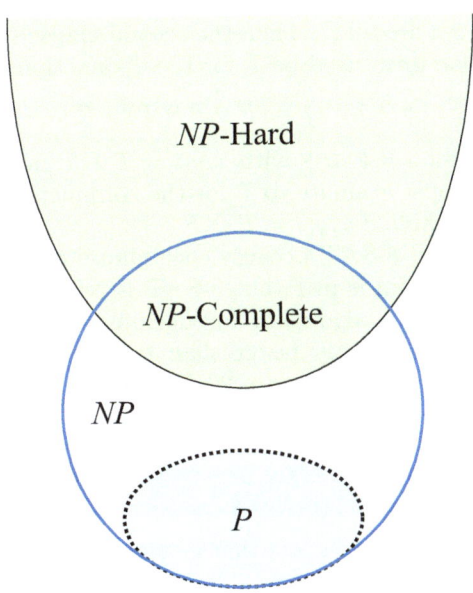

Figure 4.2 The complexity hierarchy if P ≠ NP.

complexity classes. To simplify matters further, henceforth we'll not distinguish between NP-complete and NP-hard, as the latter encompasses the former.

The SET PARTITION problem is among those problems known to be NP-hard and so intractable. There is as yet no algorithm that runs significantly faster than the 2^n try-all-possibilities algorithm. However, it is a curious situation that it has not been proven that SET PARTITION *requires* exponential time. This is (a version of) the famous Millennium Prize P =? NP problem. It is widely believed that P ≠ NP with the consequence that the NP-hard problems are indeed intractable, truly hard.

4.2.3 3-SAT Is NP-Hard

We next describe the NP-hard 3-SAT problem, variations of which are used in several origami hardness proofs, including the one we present in Section 4.3. Rather than the set of numbers used in SET PARTITION and 3-SUM, this problem is phrased in terms of logic variables, variables that can be either TRUE or FALSE, T or F for short. The variables are mixed in a logical expressions called *clauses*, with the goal deciding if there is any assignment of truth values to the variables that renders the entire logical expression T, i.e., *satisfies* the expression. Here is a 5-clause example involving three variables x, y, z:

$$(x \lor y \lor z) \land (\bar{x} \lor \bar{y} \lor z) \land (\bar{x} \lor y \lor \bar{z}) \land (x \lor \bar{y} \lor \bar{z}) \land (\bar{x} \lor \bar{y} \lor \bar{z}). \qquad (4.1)$$

The symbols ∧ and ∨ mean AND and OR, conjunction and disjunction, respectively, and a bar over a variable indicates the logical negation of the variable. Suppose x,y,z are all T. Then the second clause $(\bar{x} \vee \bar{y} \vee z)$ evaluates to T, because only one term need be T for the disjunction to be T:

$$(\bar{x} \vee \bar{y} \vee z) = (F \vee F \vee T) = T .$$

But the whole expression is FALSE with $x,y,z = T,T,T$ because the last clause is F, and all clauses must evaluate to T for the conjunction to be T. However, the expression is satisfiable if $x,y,z = T,F,F$.

The problem is called 3-SAT because each clause contains three variables. The similar 2-SAT problem is in P, but 3-SAT is NP-hard. This again means that no one has found an algorithm for deciding if an instance of 3-SAT is satisfiable that is significantly faster than trying all possibilities.[3] And if the number of variables is n, there are 2^n T/F truth assignments to check—exponentially many.

> **Exercise 4.2 [Practice] 3-SAT**
>
> Find another assignment to x,y,z that renders the expression in Equation (4.1) TRUE.

Proving a Problem Is NP-Hard The primary technique for proving that solving a problem X is NP-hard is showing that, if one could solve X quickly (in polynomial time), then one could solve some other problem known to be intractable quickly. This relies on a collection of known NP-hard problems (including SET PARTITION and 3-SAT) which can be "reduced" to problem X. Unfortunately, almost all of these reduction proofs are too complicated to present here, but we can at least give a flavor of one proof, in the next section.

4.3 Flat Folding Is NP-Hard: Proof Sketch

As we saw in Chapter 3 (Kawasaki's Theorem 3.2), we know necessary and sufficient conditions for a single vertex to be flat-foldable. Now we turn to multiple vertices. We first give some intuition why it seems difficult to determine if a crease pattern with several vertices can fold flat. Then we sketch the proof that indeed flat-foldability is NP-hard.

4.3.1 Necessary but Not Sufficient Condition

It is of course a necessary condition that if a multi-vertex crease pattern can fold flat, each of its vertices in isolation can fold flat. Now we show that this necessary condition is not always sufficient by analyzing the 3-vertex crease pattern in

---

[3] In fact there are now "SAT solvers" that can handle many instances of 3-SAT reasonably quickly, but still not in guaranteed subexponential time.

4.3. Flat Folding Is NP-Hard: Proof Sketch

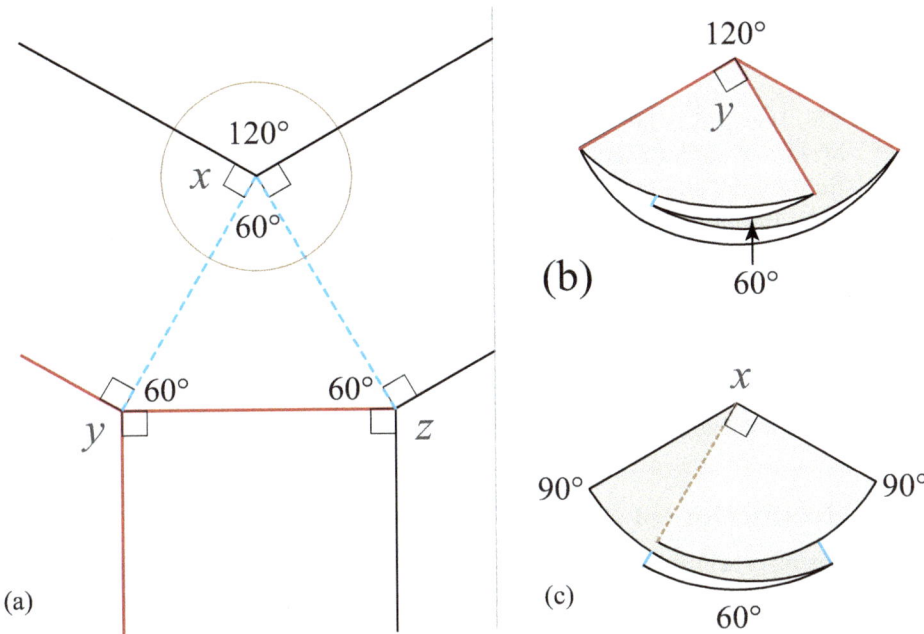

Figure 4.3 (a) Crease pattern that cannot be folded flat. (b) Flat folding of vertex y in isolation. (c) The Local-Min Lemma 3.2 is violated at vertex x. See also Figure 3.6.

Figure 4.3(a), our first triangle gadget. Let xyz represent the equilateral triangle with vertices x, y, z. First we show that each vertex can fold flat individually.

Each vertex is degree-4, and each has the same structure, sector angles $60°, 90°, 120°, 90°$. Kawasaki's Theorem 3.2 is satisfied at each vertex, odd angles = even angles = $180°$:

$$60° + 120° = 90° + 90° = 180°,$$

so we know each vertex can fold flat with some M/V assignment. A flat folding of vertex y is shown in Figure 4.3(b).

Next we argue that no M/V assignment will permit the entire pattern to fold flat.

(1) Because xyz is equilateral, its three vertices are equivalent. We focus on x without loss of generality.[4]

(2) At least one vertex must have two incident M-folds or two incident V-folds on the triangle edges. We selected in Figure 4.3(a) two V-folds incident to x, again without loss of generality.

(3) Recall the Local-Min Lemma 3.2:

---

[4] The mathematician's stock phrase "without loss of generality" means that making a particular assumption still leaves the argument general.

In any flat folding, any sector whose angle is a local min must be delimited by one mountain and one valley fold.

(4) The Local-Min Lemma fails at x: The central 60° angle is surrounded by two V-folds. Figure 4.3(c) shows an attempted folding of vertex x: Because 90° > 60° < 90°, the paper is forced to pass through itself. (The reason y can fold flat is that its 60° angle is surrounded by M- and V-folds.)

Therefore, although flat-foldability of each vertex of a multi-vertex crease pattern is a necessary condition for flat-foldability of the pattern, this example shows it is not in general a sufficient condition. We will return to this example in Section 4.3.6 where we will see that a seemingly minor change in the crease pattern permits flat-foldability.

4.3.2 Reduction of 3-SAT

We take it as established that 3-SAT (and its variations)[5] is NP-hard, and we seek to "reduce" 3-SAT to flat folding in the following sense. Given any instance of 3-SAT, we create a crease pattern that can be folded flat—assigning M or V to each crease and not creasing elsewhere—if and only if the instance of 3-SAT can be satisfied. This then will show that flat folding is NP-hard. For if the folding question could be answered quickly, then so could the 3-SAT instance, but we know from prior work that 3-SAT is NP-hard and so intractable. The theorem will then follow.

> **Theorem 4.1 Flat Folding Hard**
>
> Determining whether or not a multi-vertex crease pattern can fold flat is NP-hard.

The challenge is to construct a crease pattern that somehow behaves like 3-SAT. It is remarkable that this can be achieved, even restricting the crease pattern to use only horizontal, vertical, and 45° diagonal creases, so-called "box-pleated" patterns. We now sketch the construction at a high level.

4.3.3 Truth Values; Wires

To mimic a 3-SAT instance like that in Equation (4.1), we need to represent variables and clauses and truth values in crease patterns, and connect them according to the 3-SAT instance. Truth values are represented in **wires** that bring the variable truth assignments to the clauses. Figure 4.4(a) shows the two truth values, and Figure 4.4(b,c,d) illustrates the folding in 3D.

---

[5] The proof in the literature uses a variant called NAE 3-SAT, "Not All Equal" 3-SAT. We'll continue to call it 3-SAT.

4.3. Flat Folding Is NP-Hard: Proof Sketch

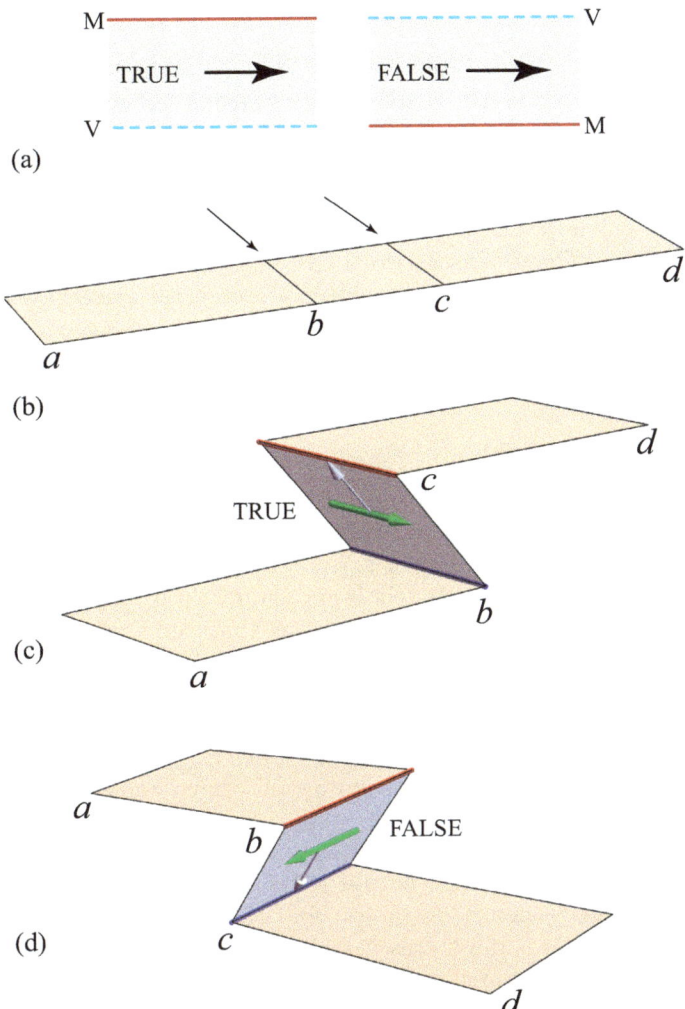

Figure 4.4 (a) T and F wires. (b) Before pleat folding. (c,d) The green arrow shows the direction of the wire. The gray arrow points toward its left.

Wires are constituted by two close parallel creases, and viewed as having a direction. Figure 4.4(b) shows the crease pattern of a wire between creases at points b and c. The creases are close enough so the two creases could not be folded both M or both V, as that would violate the Local-Min Lemma 3.2, requiring paper to cross through itself. So folding a wire flat requires one of the two possibilities of a pleat fold: Either the wire is folded so that the mountain fold is to its left and valley to its right, or vice versa. The first is identified with TRUE and the second with FALSE. See Figure 4.4(c,d). A wire then carries its variable's truth value throughout its length.

4.3.4 Gadgets

To mimic a logical expression, the crease pattern needs to have several different types of gadgets. There is a need for clause gadgets, for wire-splitter gadgets to duplicate a variable's truth value to bring it to the several clauses in which it occurs, a negation gadget to flip a truth value, a cross-over gadget to allow two wires to cross without interfering with one another, and perhaps other structures to tie it all together and ensure it "works," i.e., accurately mimics a 3-SAT instance. These constructions can be quite beautiful and intricate, essentially building a logical machine into the crease pattern. We'll detail just one gadget, a splitter gadget.

4.3.5 Splitter Gadget

As we will see (ahead in Figure 4.8), the final construction will arrange variables down the left side of the piece of paper, and clauses situated along the top side. Each variable x_i "flows" along a wire extending rightward horizontally. The truth value of x_i is "set" by choosing one of the two pleat folds at x_i's "root" on the left paper edge, and propagating that fold throughout the wire's length.

Each variable might participate in multiple clauses, and the truth value set at the root of a variable must be duplicated faithfully for each clause in which it participates. Thus there is a need for a "splitter" gadget that takes a wire as input to the left and duplicates its truth value on two emerging wires to the right.

Figure 4.5(a) shows a splitter with an input FALSE signal on the left, duplicated to two slightly thinner output wires angled off at ±45° to the right. The M/V assignments shown do indeed fold the gadget flat, as depicted in Figure 4.5(b). Figure 4.6 illustrates a folding of the input and output wires according to the M-fold of crease segment yz and the two V-folds on segments xy and xz. The total effect is to flip the triangle underneath the input wire and interchange the output wires. The conclusion is two-fold: First, the gadget folds flat with the M/V assignment shown, and second, the two output wires duplicate the truth value on the input wire; the logical signal is split.

> **Exercise 4.3 [Practice] Splitter Self-Crossing**
>
> The folding animation illustrated in Figure 4.6 has paper passing through itself between initial and final states. Argue that, for just folding Figure 4.5(a) (and not the surrounding paper), it is possible to accomplish the fold without self-crossing.

4.3. Flat Folding Is NP-Hard: Proof Sketch

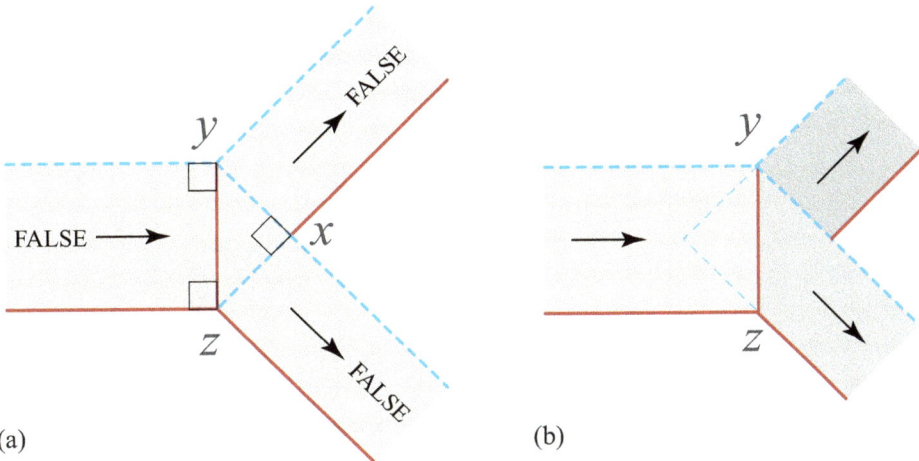

Figure 4.5 (a) Splitter gadget when input is FALSE: V-fold to left of arrow. (b) Final flat folding. The triangle xyz is now flipped under the input wire. Surrounding paper not shown.

> **Exercise 4.4 [Understanding] Splitter with Surround**
>
> Cut out and fold flat the splitter gadget in Figure 4.5(a), including the paper exterior to the gadget proper which was removed in Figure 4.5(b).

4.3.6 Two Triangles Compared

We return to the argument that the crease pattern in Figure 4.3 cannot fold flat, and compare it to the flat folding in Figure 4.5(b). The two crease patterns differ only in the shapes of the triangles. The central triangle xyz in Figure 4.3 is equilateral, whereas the triangle in Figure 4.5 is a right isosceles triangle, angles $45° - 45° - 90°$.

As before, Kawasaki's Theorem 3.2 is satisfied at each vertex of the right isosceles triangle:

$$x : 90° + 90° = 90° + 90°$$
$$y : 90° + 90° = 45° + 135°.$$

So each vertex can fold flat in isolation. We next parallel the argument in Section 4.3.1.

(1) Because xyz is right isosceles rather than equilateral, we need to consider two cases: x, and y (y and z are equivalent).

(2) At least one vertex must have two incident M or two incident V creases along triangle edges.

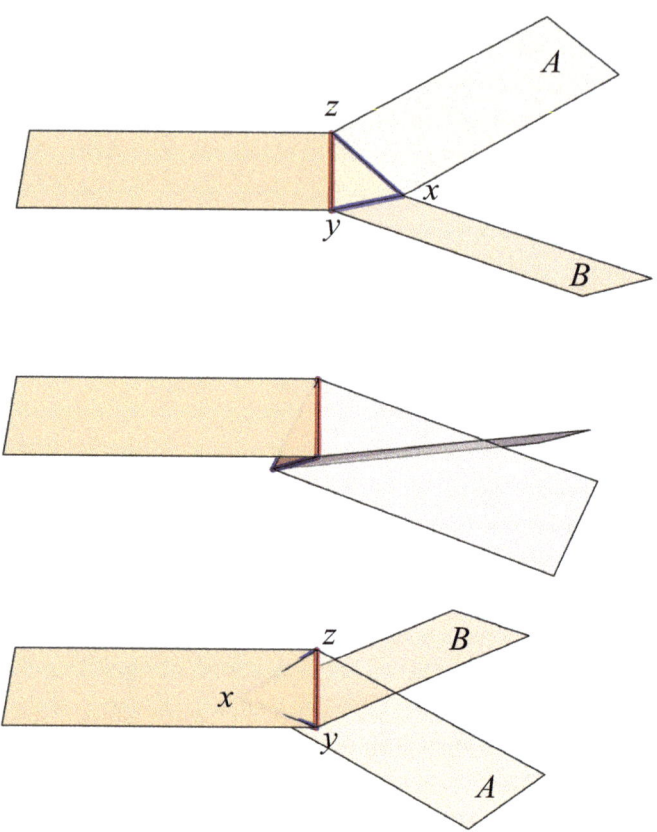

Figure 4.6 Snapshots of a splitter triangle xyz folding; x flips underneath and wires A and B swap positions. Animation: https://cs.smith.edu/~jorourke/MathOrigami/.

(3) Two cases.

- a. First assume two V creases are incident to vertex y as in Figure 4.7(b). Then the Local-Min Lemma 3.2 fails: $90° > 45° < 90°$. So the M/V assignment shown surrounding y fails to fold flat.

- b. Second, instead assume two V creases are incident to vertex x. As Figure 4.7(c) shows, then the Local-Min Lemma is satisfied at both x, where $90° = 90° = 90°$, and at y because there are M and V creases surrounding the central $45°$ angle. And indeed Figure 4.7(c) follows the flat folding in Figure 4.5(b).

We will encounter our third triangle-based origami gadget in Section 4.5.2.

4.3. Flat Folding Is NP-Hard: Proof Sketch

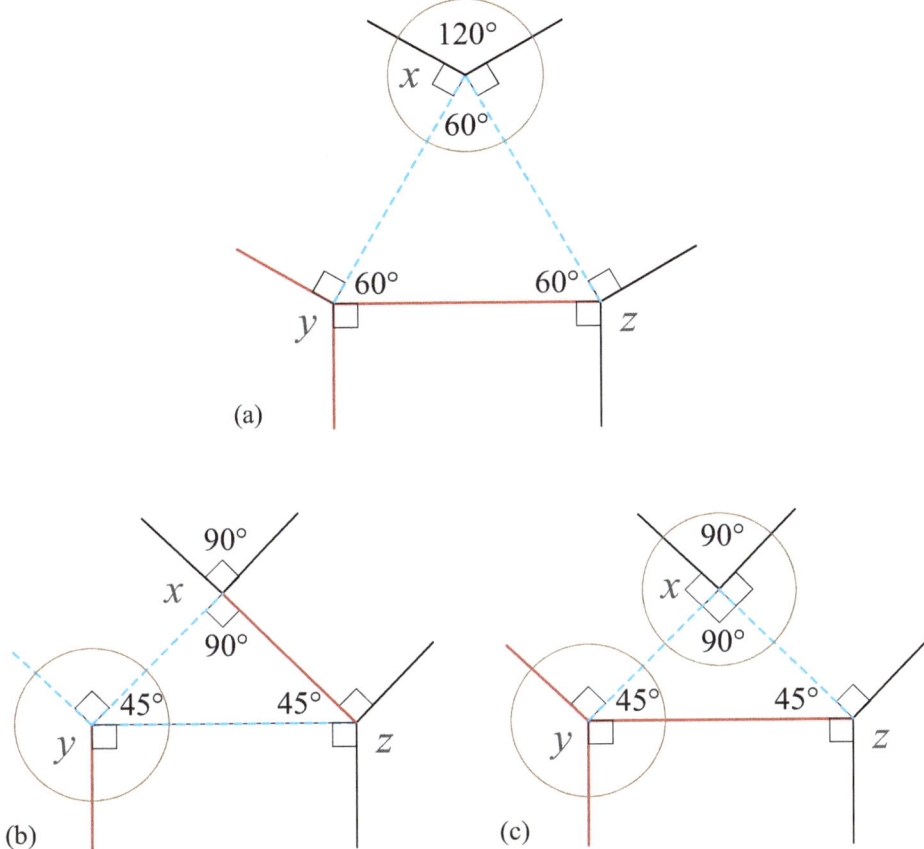

Figure 4.7 (a) A repetition of Figure 4.3(a). (b) Two V-folds incident to y: cannot fold flat. (c) Two V-folds incident to x: can fold flat, as in Figure 4.6.

4.3.7 Complications

Two characteristics of the splitter gadget require attention in the complete design. First, the output wires are thinner (by a factor of $1/\sqrt{2}$) than the input wire, a situation that cannot be permitted to propagate throughout the design. Second, the output wires are angled at 45° with respect to the horizontal. The proof solves both problems by treating output wires as input wires at the further stages, both thickening and turning them.

The design requires wires to cross without interfering with each others' truth signals. This requires a tricky cross-over gadget. In fact the original proof contained a flaw in the cross-over gadget that was not detected for 20 years. The turning of wires 45° permits arranging that all wire crossings occur at right angles, considerably simplifying the cross-over gadget.

Splitting wires sometimes results in **noise wires**, wires created but not needed to feed into a clause. These can be arranged to run off the boundary of the paper rectangle.

4.3.8 The Whole Shebang

Considerable care is needed to connect all the various gadgets together to faithfully simulate a given 3-SAT variant problem.

A rough idea of a full construction is displayed in Figure 4.8, showing four variables x_1, x_2, x_3, x_4 along the left border, and two clauses along the top.[6] For example, the x_1 wire reaches the first clause gadget, but bypasses the second clause gadget (and then runs off the right end) because it is not one of the three variables participating in that second clause. The x_4 wire skips the first clause but enters the second as $\bar{x}_4$.

If the crease pattern can fold flat, then such a folding will set each variable to T or F, and propagate those values to the clauses, whose flat folding guarantees satisfiability. In the reverse direction, if the classes can be satisfied by a specific set of truth values, then M/V setting the variables accordingly folds the construction flat.

The conclusion is that the whole construction folds flat if and only if the logical expression from which it is derived is satisfiable. And because it is known that satisfiability is NP-hard, this establishes that flat folding must also be NP-hard: Theorem 4.1.

> **Open Problem 4.1 Orthogonally Aligned Creases**
>
> We've seen that Theorem 4.1 holds even restricted to "box-pleating" crease patterns, restricted to angles that are multiples of $\pm 45°$ (evident in Figure 4.8). The complexity is unknown if creases are restricted to $\pm 90°$.

4.4 Turing-Completeness

We've just seen (Theorem 4.1) that deciding whether a pattern can fold flat is a difficult question. It could be decided by trying all M/V assignments, and all layering orderings, but that is infeasible for even a mildly complex crease pattern. In this section we add to the evidence that flat folding is intricate by showing that flat folding is "Turing-complete." Explaining this will require another excursion into theoretical computer science before we reach the crease-pattern construction that establishes the claim. We'll see again the central role of a different type of triangle gadget (Section 4.5.2). We start the excursion with Turing machines.

---

[6] As mentioned earlier, these are NAE 3-SAT clauses.

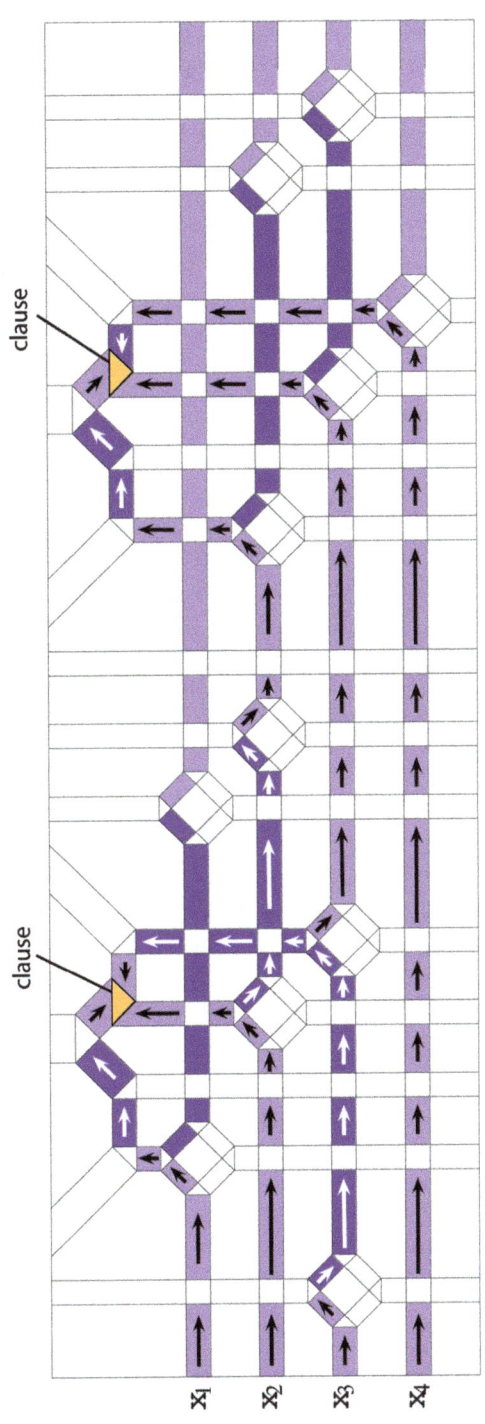

Figure 4.8 Left clause: $x_1 \lor x_2 \lor x_3$. Right clause: $x_2 \lor x_3 \lor \bar{x}_4$. Dark arrows T, white arrows F. [From Hull (2020). Reprinted by permission of Cambridge University Press.]

4.4.1 Turing Machines

A ***Turing machine*** is an abstract computer invented by Alan Turing in 1936, well before the first digital computers were constructed in the mid 1940s. A Turing machine can make only the most elementary operations: for example, reading or writing a symbol on a "tape." His motivation was to define the simplest possible model of what is computable. Since then his definition has been accepted: A Turing machine can compute anything computable, converting input data to outputs by running a program (in any programming language, e.g., Python or Javascript). So a program that takes as input a whole number x and returns its cube, x^3, can be computed on a Turing machine. Similarly a program that takes as input your financial data for the calendar year and calculates how much you owe in income tax, could be computed on a Turing machine.

4.4.2 Turing-Complete

A device is ***Turing-complete*** if it can perform any computation that could be programmed on a Turing machine, which means it could be programmed on any computer. This reflects the power or complexity of the device. A calculator or a digital clock is not Turing-complete because neither can perform general-purpose computation: They both have limited logic capabilities.[7] The result in this section shows that a crease pattern can be constructed so that flat folding it performs a computation—any computation—i.e., flat folding is Turing-complete. Theoretically you could figure out your income tax by folding paper!

The path to this result goes through two conceptual devices each fascinating in their own right: the two-dimensional game of LIFE, and one-dimensional cellular automata.

4.4.3 The Game of LIFE

The mathematician John Conway invented in 1970 a game he called LIFE played on an infinite grid of squares. It is not really a game, but rather a "cellular automaton" that evolves from an initial configuration. Each square cell may either be alive/occupied (1), or empty/dead (0). At each time step, each cell is updated according the configuration of its eight immediate neighbors:

- A live cell dies in the next generation if it has $0, 1, 4, 5, 6, 7, 8$ live neighbors, but lives on if it has $2, 3$ live neighbors.
- A dead cell becomes alive if it has exactly 3 live neighbors, and otherwise remains dead.

These simple rules create an amazing variety of evolving patterns which burdened the world's primitive computers when Martin Gardner wrote his first *Scientific American* column on the topic. Conway's hand-explorations led him

---

[7] Even though today both are implemented on phones which do have general-purpose computational abilities.

4.4. Turing-Completeness   49

to pose the question (and offer $50 for resolution): Is there is any starting pattern that grows live cells without bound? This was soon answered positively by Bill Gospers with his remarkable glider gun shown in Figure 4.9. The blue configuration constitutes the "gun," and the red 5-cell clusters are "gliders," emitted by the gun every 30 time steps, which glide off diagonally without end.

With this tool, it was not long before Conway conjectured and later proved (in 1982) that LIFE is Turing-complete. Conway liked to say this means some starting pattern could calculate the digits of π as it evolves. And in fact years later enthusiasts designed a pattern that indeed "prints" the digits of π in the infinite grid of cells: see Figure 4.10. Conway's proof uses gliders to form the basic AND, OR, NOT logic gates that power any computer.

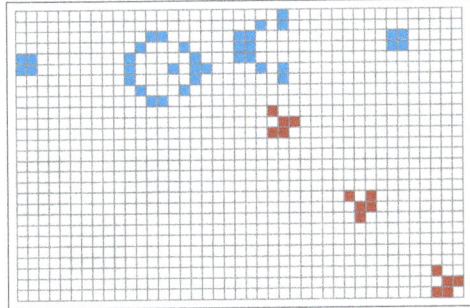

Figure 4.9 Gosper's glider gun (blue) after emitting three South-East headed gliders (red).

Figure 4.10 Output of the LIFE π-calculator, designed by Adam P. Goucher, in collaboration with David Greene and Paul Chapman.

4.4.4 1D Cellular Automata

The game of LIFE is a 2D *cellular automaton*, 2D because the grid of squares fills the plane. A 1D (one-dimensional) cellular automaton consists of a single infinite row of cells. As in LIFE, the cells can be live (1) or dead (0) and evolve according to eight rules, one rule for each pattern of a cell and its left and right neighbors. For example, the rule shown in Figure 4.11 means that a dead cell surrounded by live cells to both sides comes to life in the next generation. The generations are typically displayed horizontally below one another, which highlights the evolution of a pattern over time.

4.4.5 Rule 110

One set of eight rules, known collectively as Rule 110, was extensively studied by Stephen Wolfram (the creator the MATHEMATICA software that produced the 3D figures in this book). The eight rules are displayed in Figure 4.12(a), as well as the evolution starting from a single live cell in Figure 4.12(b).

Figure 4.11 One of the eight rules in Rule 110. A live cell is indicated by 1, dead by 0, displayed as black and white respectively.

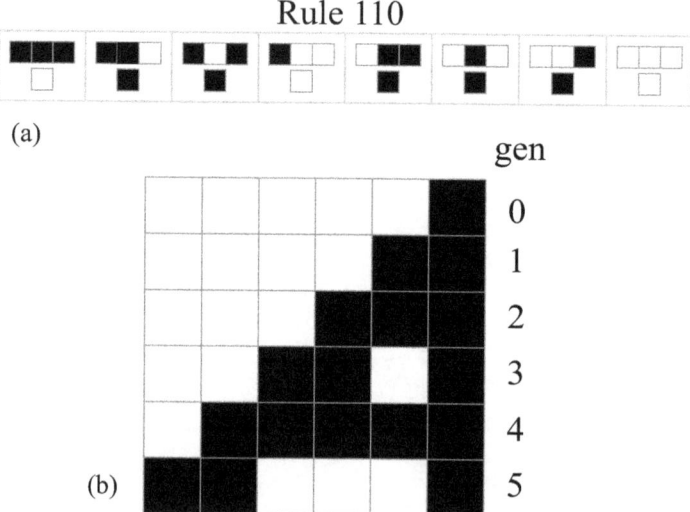

Figure 4.12 (a) Rule 110: Eight rules, one for each pattern of three cells. (b) Five generations evolving from a single live cell in generation 0.

4.5. Flat Folding Is Turing-Complete: Proof Sketch

In words, Rule 110 states that a live cell remains live if it has 0 or 1 live neighbors, and a dead cell comes to life if its right neighbor is alive. Despite the simplicity of the rules, the evolution is quite complex, with patterns replicating and "moving" through the generations in a manner reminiscent of Gosper's gliders. The complexity of pattern evolution under Rule 110 led Wolfram to conjecture that Rule 110 is Turing-complete, a conjecture proved by Matthew Cook in 1994.

> **Exercise 4.5 [Understanding] Rule 110**
>
> Using the Rule 110 rules in Figure 4.12(a), calculate five generations of evolution starting from 1 0 1.

A specific computation under Rule 110 is effected by filling the first row with a pattern tailored to the computation, and letting the pattern evolve according to the rules. A final pattern, appropriately interpreted, constitutes the output of the computation.[8] Thus Rule 110 is Turing-complete.

4.5 Flat Folding Is Turing-Complete: Proof Sketch

We now sketch the proof of the claim:

> **Theorem 4.2 Flat Folding is Turing-Complete**
>
> Flat folding a crease pattern (allowing optional creases) is Turing-complete.

The main idea is to simulate Rule 110 by a crease pattern on a half-infinite piece of paper. Simulation of the eight individual rules of Rule 110 is accomplished via logic-gate gadgets, implementing logical connectives operating on truth values TRUE or FALSE propagated down wires. A typical logic gate is that for conjunction: AND, symbolically $\wedge$ as used in Section 4.2.3. An AND-gate is a gadget with two input wires and one output wire, which outputs TRUE when both inputs are TRUE, and outputs FALSE otherwise. All the rules of Rule 110 can be simulated by several logic-gate gadgets: AND, OR, NOT, and the less common NAND and NOR. These are similar to the gadgets described in Section 4.3. For example, using live $= T =$ TRUE and dead $= F =$ FALSE, the rule in Figure 4.11 could be expressed as

$$T \wedge F \wedge T \to T.$$

---

[8] Here I am ignoring the details of Cook's proof, which emulates a "cyclic tag system."

Unlike the NP-hard proof, the piece of paper fills an infinite half-plane, with the horizontal upper boundary simulating the starting row of cells of a 1D cellular automaton. Setting the specific pattern of the start cells to perform a computation is accomplished by flat folding the crease pattern. The flat folding along the top boundary of the paper ripples truth values downward, through the logic gates simulating Rule 110. A correct simulation guarantees Turing-completness, relying on knowing that Rule 110 is itself Turing-complete.

Note that there is no longer a question of whether or not the crease pattern can fold flat. It always can fold flat. It is the act of folding it flat that accomplishes the computation indicated by its start cells.

4.5.1 Wires

As in the NP-hard proof in Section 4.3, truth value signals TRUE and FALSE are propagated down directed *wires*, patterns of creases that only permit one of two flat foldings. Recall from Figure 4.4 that wires in the NP-hard proof were two parallel creases close enough so that their folding assignments had to be MV or VM (to satisfy the Local-Min Lemma 3.2), representing the two truth values. In the Turing-completeness proof, a wire consists of three parallel creases: a mandatory M-fold of the middle crease, and two optional V-folds along creases to either side, as depicted in Figure 4.13(a). Here, as in the SET PARTITION example in Section 4.2.1, an optional crease could be left uncreased or *inactive*. Otherwise it is *active*. If the pleat is folded to the right relative to its direction then it is TRUE; if it is folded to the left then it is FALSE. In order for a wire to have a well-defined truth value, it must be the case that only one of its optional valley creases is active. Although it is possible to fold both optional valley folds in the local vicinity of a wire, the interaction between a wire and the gadget to which it connects ensures that only one of the TRUE and FALSE folds permits flattening the gadget. Although this can be proven for all the gadgets in the construction, the conclusion can be reached for the one particular gadget we'll study in the next section via a simple argument based on Maekewa's Theorem 3.1.

Comparison of Figure 4.4 and Figure 4.13 shows that the pleat foldings along a wire end up being the same in the two constructions. So it may be mysterious why the NP-hard wires have two parallel creases but the Turing-complete wires have three parallel creases. The reduction from 3-SAT requires the flat folding to enforce a specific M/V folding of the creases at the sources of the variables (down the paper left side in Figure 4.8) to mimic a specific satisfying truth assignment. In contrast, the Turing-complete construction is a "machine" that needs to permit the logic variables (along the paper top row) to choose specific T/F assignments to the variables for one computation and different T/F assignments for another computation. So there must be the option of choosing either T or F, i.e., folding the center M-fold to the right or to the left.

As in the NP-hard proof, gadgets are needed for various purposes: for implementing the logic gates, for reorienting wires carrying truth signals,

4.5. Flat Folding Is Turing-Complete: Proof Sketch

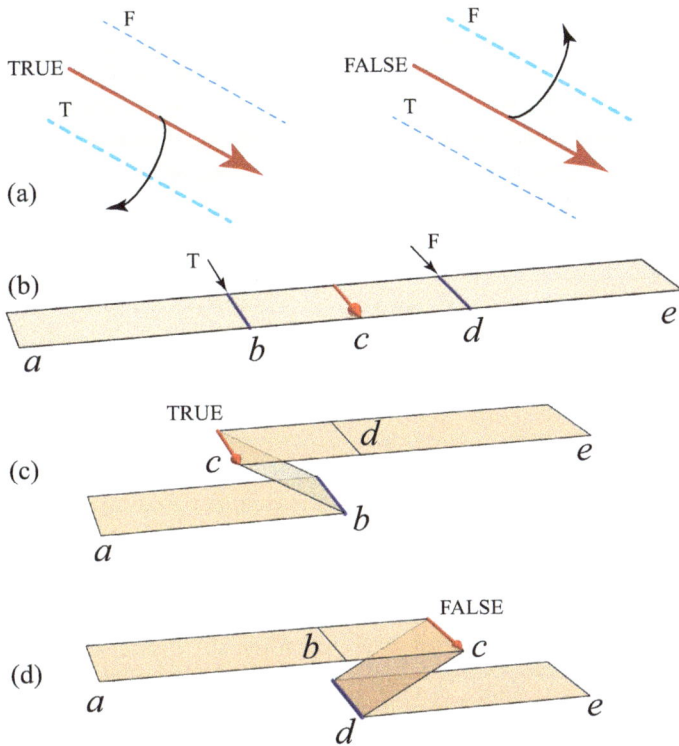

Figure 4.13 (a) Either the T crease is active and the F inactive, or vice versa. (b) Crease pattern before folding. (c,d) TRUE and FALSE pleats.

crossing gadgets, splitting gadgets, even "eater" gadgets to absorb unneeded extra "noise" wires, as such wires cannot run off the boundary of the infinite half-plane. And all these gadgets must work correctly based on the three-parallel-crease wires shown in Figure 4.13(a). We opt to detail just one gadget, again a splitter gadget.

4.5.2 Splitter Gadget via Triangle Twist

We next explain the "triangle twist" gadget, Figure 4.14(a), which serves several purposes: It propagates wire signals from vertical to $\pm 120°$ with respect to the vertical, and it splits and negates the input truth value. To split and duplicate the input truth value (as does the splitter in Figure 4.5, used in the NP-hard proof) requires supplementing the gadget by NOT-gates on the output wires to negate the truth value, gates which we do not show.

This is our third gadget based on a central triangle. Figure 4.14(a) differs from the splitter gadget in Figure 4.5 in that the central triangle is equilateral, like that in Figure 4.3. But the input and output creases to the equilateral triangle are arranged differently than those in Figure 4.3.

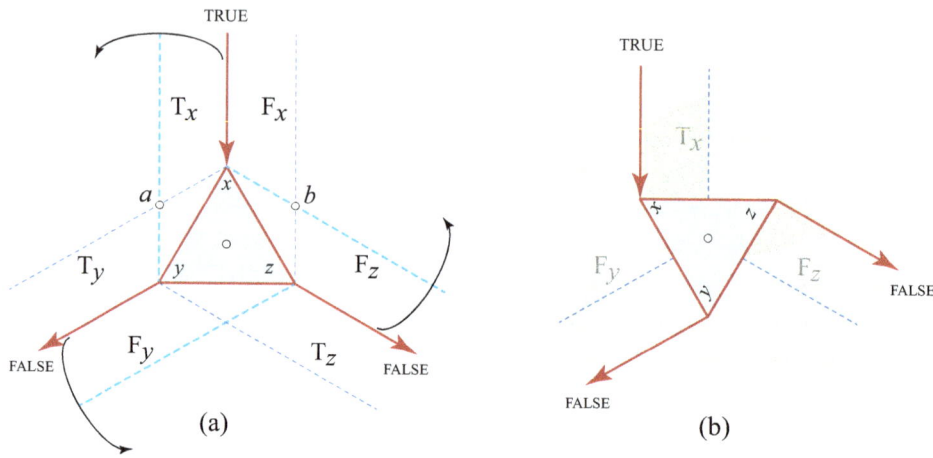

Figure 4.14 (a) Crease pattern. Arrows indicate pleat folding over T or F valley folds. Optionally unfolded V creases lightly shaded. (b) Folded flat. Now the triangle is twisted 60° counterclockwise and T/F creases are underneath.

Figure 4.14(a) shows the crease patterns when the input to the apex x of the the triangle is TRUE. Because the input is TRUE, the central vertical mountain crease valley-folds over the T optional crease, leaving the F optional crease uncreased. So Figure 4.14(b) shows the T fold underneath. Continuing with the folding, the mountain creases emanating from vertices y and z both fold over the F valley creases, and so create FALSE signals on the output wires.

Note that in the flat folding in Figure 4.14(b), the triangle is twisted 60° from its original pattern. A sense of the how the triangle gets twisted while folding flat can be seen in the snapshots in Figure 4.15. These snapshots (from an animation) only give a sense of the dynamics, because, first, the paper outside the wire channels is not included, and second, as we will see in Chapter 5 (Section 5.4.4), the folding requires bending the paper, which is not simulated in the animation.

> **Exercise 4.6 [Understanding] Triangle Twist**
>
> Cut out and fold Figure 4.14(a) when the input is FALSE rather than TRUE as shown.

> **Box 4.1 Proof by Contradiction**
>
> A *proof by contradiction* is a proof that assumes the claim is FALSE and from this assumption derives a contradiction, showing that in fact the claim must be TRUE.

4.5. Flat Folding Is Turing-Complete: Proof Sketch

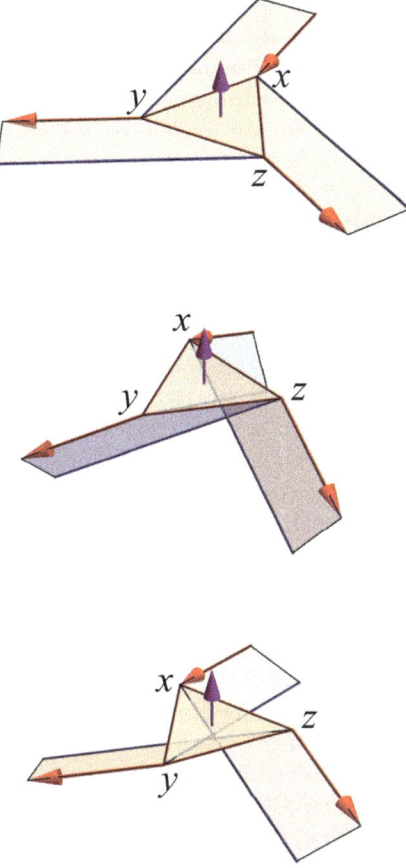

Figure 4.15 Snapshots of a triangle twist gadget folding. Animation: https://cs.smith.edu/~jorourke/MathOrigami/.

Not Both Optional Creases Next we prove that the input wire to triangle vertex x in the splitter gadget can have only one of its two optional valley creases folded/active, as a result of the interactions between the wires and the need to fold the gadget flat. First we need some notation. Let T_x and F_x be the two optional V-creases to the right and left of the mandatory M-crease incident to x. Label T_y, F_y, T_z, F_z similarly; see Figure 4.14(a).

Suppose, in contradiction to the claim, that both both optional creases are folded, that is, both T_x and F_x are V-folded. (See Box 4.1 on proofs by contradiction.) Then T_y must be uncreased, i.e., T_y cannot be V-folded, because if it were, then four V-creases would be incident to the intersection point labeled $a = T_x \cap T_y$ in the figure. This violates Maekawa's Theorem 3.1. Similarly, F_z cannot be V-folded, because then four V-creases would be incident to point $b = F_x \cap F_z$.

This leaves vertex x of degree-3: the input mountain crease and the triangle segments yx and zx, while T_y and F_z are uncreased. But a degree-3 vertex cannot satisfy Maekawa's Theorem 3.1 (which implies the Even-Degree Lemma 3.1). This contradiction establishes the claim: A wire allows only one of the two V creases to be active.

4.5.3 The Whole Shebang

Figure 4.16 shows an amazingly intricate schematic of the crease pattern that simulates a single cell in the Rule 110 automaton. The three vertical inputs along the top boundary of the paper correspond to a cell and its two neighbors, labeled A,B,C and colored blue, red, green respectively. The central purple vertical wire near the bottom is the output, setting the next generation of the B cell. The actual crease pattern is not illustrated, instead symbols and labels and their connections are shown, annotated by colors. For example, the pale yellow triangles are triangle twist gadgets, white hexagons are NAND or NOR or OR gates, triangles labeled E are eater gadgets absorbing gray noise wires, and so on.

One of the many differences between the NP-hard construction and the Turing-complete construction is that the former is focused on a given instance of 3-SAT: The designed crease pattern will fold flat only if that particular instance is satisfiable. Whereas the Turing-complete construction is a general-purpose device that can perform any computation by appropriate settings of the top-row inputs.

A natural concern is the need to start with an infinite piece of paper, another contrast to the NP-hard construction. However, both the 1D cellular automaton and the 2D game of LIFE assume an unbounded collection of empty cells, analogous to a Turing machine's unbounded blank "tape." But all these devices only employ a finite number of cells for their computations.

Of course, actually computing with this origami machine is quite impractical! Nevertheless, it is—remarkably—theoretically possible.

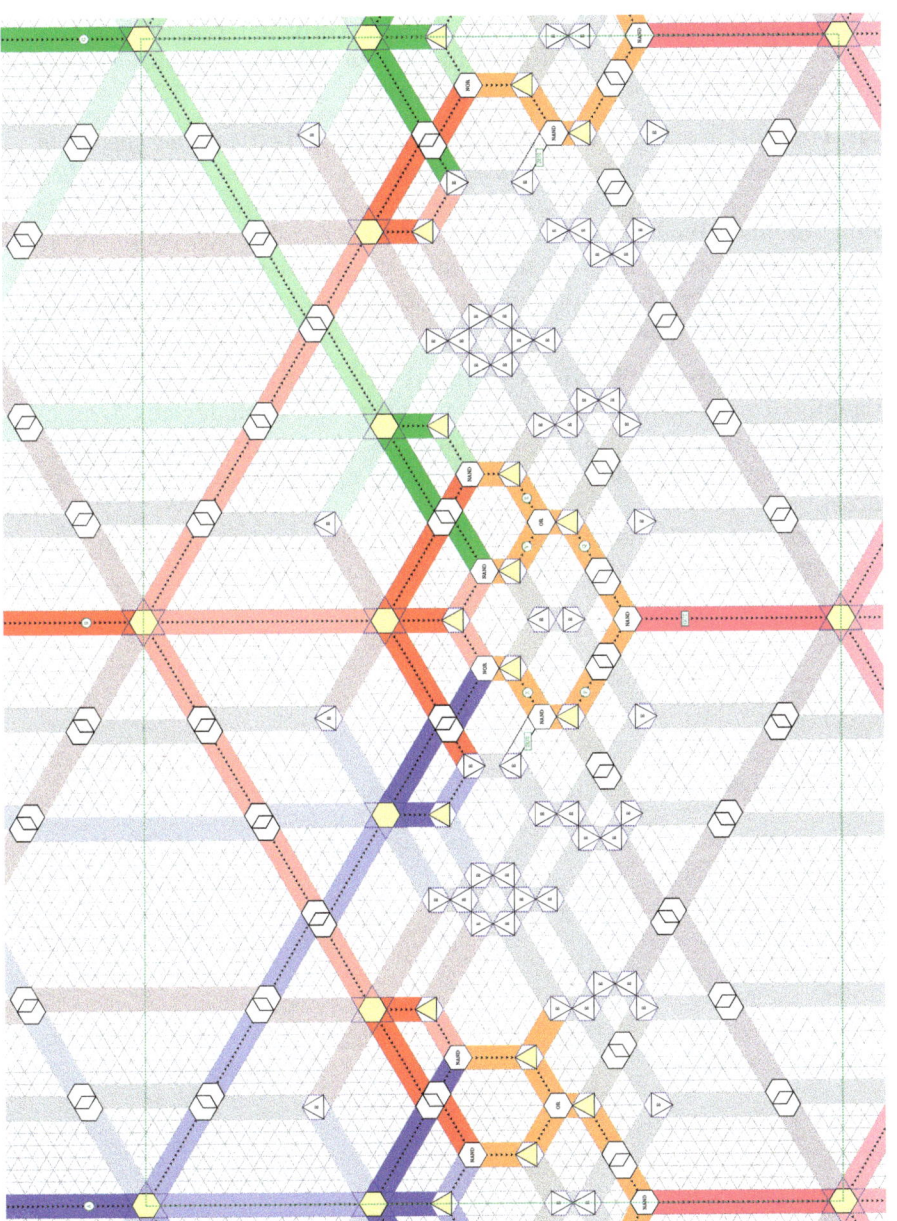

Figure 4.16 Construction for three input cells at top (blue, red, green), and one output cell (purple) at middle-bottom. [From Hull and Zakharevich (2023). Reused by permission of the authors. All rights reserved.]

4.6 Technical Notes

Sec. 4.2: P, NP-Complete, NP-Hard The best place to learn about computational complexity is an undergraduate algorithms textbook, such as Cormen et al. (2022, Ch. 36).

Sec. 4.3: Flat Folding Is NP-Hard: Proof Sketch The original proof that flat folding is NP-hard appeared in Bern and Hayes (1996). A multi-author collaboration, Akitaya et al. (2016), corrected the Bern–Hayes proof, and extended it to "box-pleating" crease patterns. My presentation relies heavily on Hull's *Origametry* book (Hull, 2020, Thm. 6.21).

Sec. 4.4: Turing-Completeness Any undergraduate text on the theory of computation is a source for Turing machines, such as (Sipser 2012). To explore Conway's game of LIFE interactively, visit https://conwaylife.com/. Stephen Wolfram's *A New Kind of Science* (Wolfram 2002) covers 1D cellular automata in great detail, and the universality of Rule 110 in particular. Mathew Cook's original proof that Rule 110 is Turing-complete appeared in Cook (2004).

Sec. 4.5: Flat Folding Is Turing-Complete: Proof Sketch That "flat origami is Turing-complete" was proven by Hull and Zakharevich (2023) in a paper with that title. My presentation relies almost solely on this paper. There is also a nice *Quanta Magazine* article on the proof by Cepelewicz (2024).

Subsequent work showed that the need for optional creases and infinite paper in Theorem 4.2 can be circumvented (Eppstein 2024).

5
Rigid Origami and Degree-4 Vertices

5.1 Introduction

In this chapter we introduce *rigid origami*: origami of flat rigid plates hinged along shared edges. Rigid origami significantly restricts what can be folded—for example, folding the standard origami crane requires bending of paper, as do most origami designs. But rigid origami can create useful structures, useful in engineering (deployable structures, i.e., unfolding solar arrays in outer space), kinetic architecture, and even furniture design. In rigid origami, the final folded shape is often less interesting than the dynamics of the configurations that reach that final shape.

In this chapter we study rigid origami via crease patterns comprised of degree-4 vertices, for which there has been recent significant advances in mathematical understanding. In particular, we will emphasize the remarkable half-tangent relationships discovered that the dynamics of all flat-foldable degree-4 vertices obey. We analyze five examples: an open box, the "plus-sign"—a warm-up for the Miura map fold which we analyze in some detail—the square twist, and the flexible half-octahedron. Finally, we will sketch a proof that rigid origami is NP-hard.

5.2 Rigid Origami

> **Box 5.1 Faces vs. Facets**
>
> The contiguous regions of a crease pattern, delimited by crease segments or boundary edges, are called *faces* or *facets*. Although the term "facets" is often used in the origami literature, both graph theory and polyhedral geometry use "faces," with geometry reserving "facets" for higher-dimensional faces.

5.2.1 Open Box Example

We start with a simple example, which we describe only qualitatively to give a sense of rigid folding. Consider the crease pattern drawn on an octagon in Figure 5.1(a). The five squares have side length 1, so the four red diagonals each have length $\sqrt{2}/2$. These four (red) diagonals are M-folds; all the remaining internal edges (blue) are V-folds. In rigid origami, the faces of a crease pattern (see Box 5.1) cannot flex or bend. The only bending permitted is hinging along crease segments shared by two faces. Each face is rigid, as if made of metal. Each internal crease segment is a hinge with full 360° dihedral angle range, fold angle from 0° (flat, uncreased) to fold angles ±180°.

Figure 5.1(b,c) shows one possible folding progression, reaching a (lidless) cube with four doubled triangle "fins" aiming toward the cube center. Each fin then rotates toward a cube face in Figure 5.1(d) until it merges flush with a face in Figure 5.1(e,f).

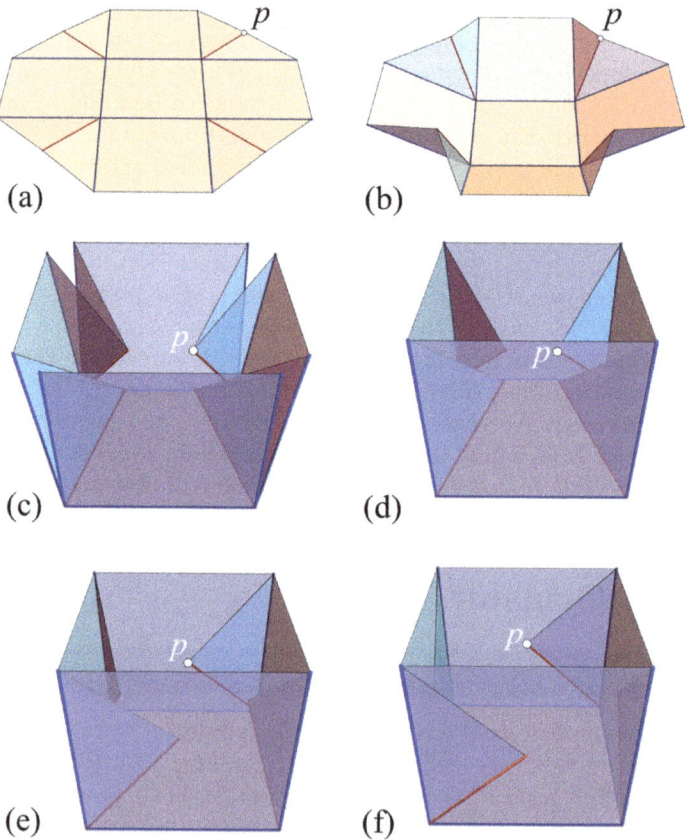

Figure 5.1 Rigid folding of an open box, tracking one point p. Animation: https://cs.smith.edu/~jorourke/MathOrigami/.

5.2. Rigid Origami

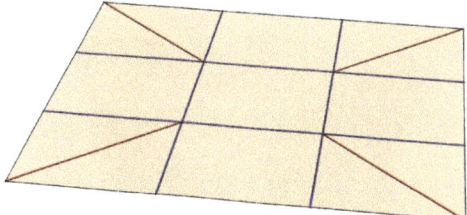

Figure 5.2 Crease pattern on a 3 × 3 square.

This example illustrates two points. First, note that each of the four vertices is degree-5, four V-folds and one M-fold. So Maekawa's Theorem 3.1 is not satisfied. But that theorem concerns flat foldings, whereas here the final folded state is a 3D non-flat structure. Our focus in this chapter will differ from this example in that all vertices will have degree-4, and Maekawa's Theorem will be satisfied and flat states will play a role.

Second, the rigid face plates do not cross through one another during the motion. They can lay flush on another face, as do layers of faces in a flat folding, but not penetrate. Noncrossing is a natural restriction for the many practical rigid origami constructions. Exercise 5.1 explores noncrossing of faces further.

> **Exercise 5.1 [Challenge] Cube from 3 × 3 Square**
>
> (a) Suppose instead of an octagon, the same crease pattern is based on a 3 × 3 square, as in Figure 5.2. Argue that, if the same folding sequence as in Figure 5.1 is followed, faces will collide and cross into one another.
>
> (b) Nevertheless, show that there is a different folding sequence that avoids face–face collisions.

We next describe a few engineering applications.

5.2.2 Folding Furniture

Folding furniture for storage is common and not in need of origami design principles. But there are artistic designs that are inspired by rigid origami. An example is Brian Ignaut's delightful folding "Loop Table": Figure 5.3. Its 10 hardwood panels are hinged so that two opposite twists unfold the structure from a flat loop to a side table in an elegant motion. The reader is welcome to build a paper model following Figure 5.4 to experience the motion.

Perhaps not surprisingly, the artist behind this Loop Table was the lead solar array designer at SpaceX prior to turning to furniture. Solar arrays are the premier application of rigid origami.

Figure 5.3 Loop Table. [Reprinted by permission of the artist, Brian Ignaut. https://enjoydof.com/.] Animation: https://cs.smith.edu/~jorourke/MathOrigami/.

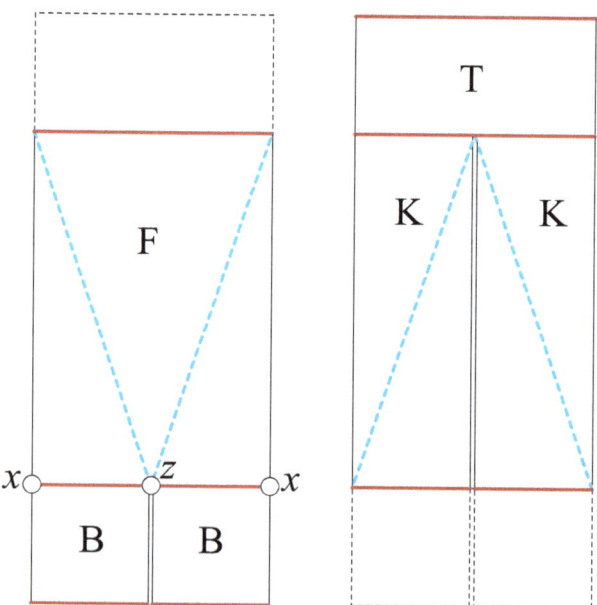

Figure 5.4 Loop Table template. F,K,T,B = Front, bacK, Top, Bottom. The two points x on the base squares B rotate about z to join together.

5.2. Rigid Origami

Figure 5.5 *James Webb Space Telescope* mirror. Three hexagonal mirrors to either side of the blue paths were rotated into place. [Credit: National Aeronautics and Space Administration.]

5.2.3 Solar Arrays

Solar arrays are formed of rigid photovoltaic cells, which must fold to fit inside a rocket's storage bay, and then unfold in space to a large area to capture energy from the sun. These needs all point to rigid origami.

The same rigidity was needed for the *James Webb Space Telescope* primary mirror, which was folded on launch and unfolded in space. However, the folding was rather simple: Of the 18 gold-plated hexagonal mirrors that comprise the full mirror, a stack of three on the left and three on the right folded from storage behind to create the full mirror. See Figure 5.5.

A more origami-intensive design, named HanaFlex ("Hana" is Japanese for "flower"), is shown in Figure 5.6. The design is based on the origami *flasher* pattern studied by several origamists since the 1960s. Famed origamist Robert Lang was part of the design team. The crease pattern for one version is illustrated in Figure 5.7(a). The creases appear curved, but in fact are gentle turnings of straight creases emanating from the central hexagon like a camera shutter. When the pattern is fully folded, as in Figure 5.7(b), it forms a hexagonal cylinder intended to stow in the rocket.

Although the HanaFlex design has not yet been deployed into space, the *Japanese Space Flyer Unit* satellite did dock with the Space Shuttle *Endeavour* in 1996. Its solar panels extended an impressive 24 meters once unfurled in space: see Figure 5.8. The rigid origami design behind the unfurling was invented by Koryo Miura, the basics of which design is the focus of the next section.

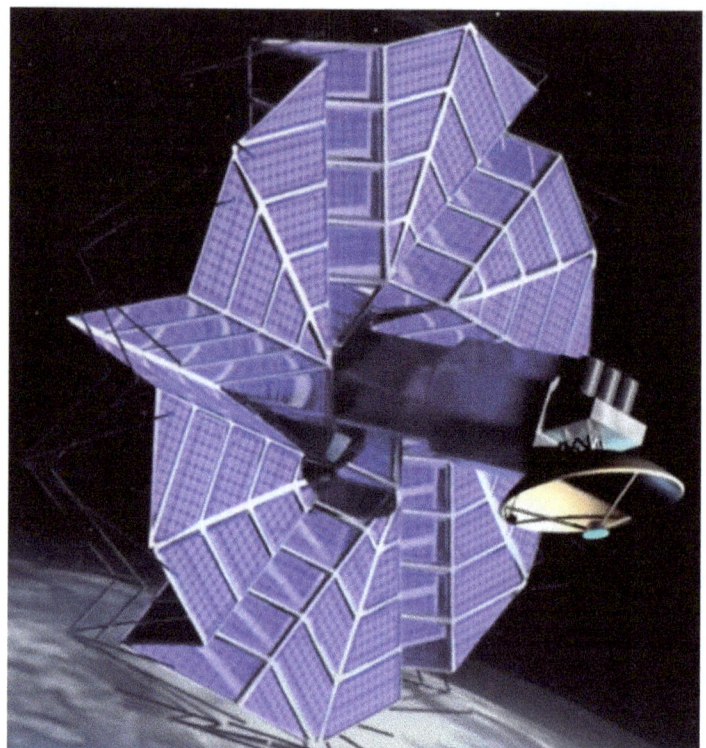

Figure 5.6 HanaFlex solar array. [Reprinted by permission of the authors, Larry Howell and Spencer Magleby (Brigham Young University).]

5.3 Miura Map Fold

We emphasized in Section 2.5 the difficulty of folding a rectangular map flat when given an arbitrary M/V pattern. But a particular M/V pattern makes it easy to fold maps, so easy that many map manufacturers follow this pattern in their design: Figure 5.9.

The particular design is called the ***Miura-ori*** pattern ("ori" meaning "folding"), or the Miura map folding. As mentioned, it was invented by the Japanese astrophysicist Koryo Miura for the purpose of rigidly unfolding a satellite solar array in space. He invented the folding in the 1970s but had to wait to 1995 for it to be launched in a rocket. The design is space efficient, as we'll see (Exercise 5.2).

Not only is the Miura pattern rigidly foldable, it has a single ***degree-of-freedom***, 1-DOF for short. This means that folding any single crease in the pattern causes all the creases to fold in concert. This is in contrast to the box example in Section 5.2.1, which has many degrees of freedom: portions of the design that can fold independently of other portions.

5.3. *Miura Map Fold*

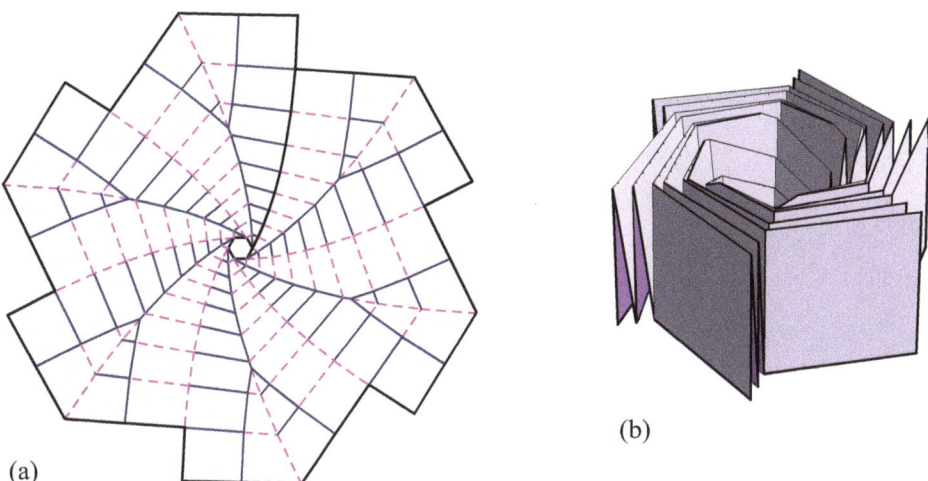

(a)

(b)

Figure 5.7 (a) Flasher crease pattern. (b) Fully folded. [From Lang et al. (2016). Used with permission of the American Society of Mechanical Engineers, from Lang et al. (2016); permission conveyed through Copyright Clearance Center, Inc.]

Figure 5.8 The *Japanese Space Flyer Unit* with solar array unfurled. Photographed from the *Endeavour*. [Credit: National Aeronautics and Space Administration.]

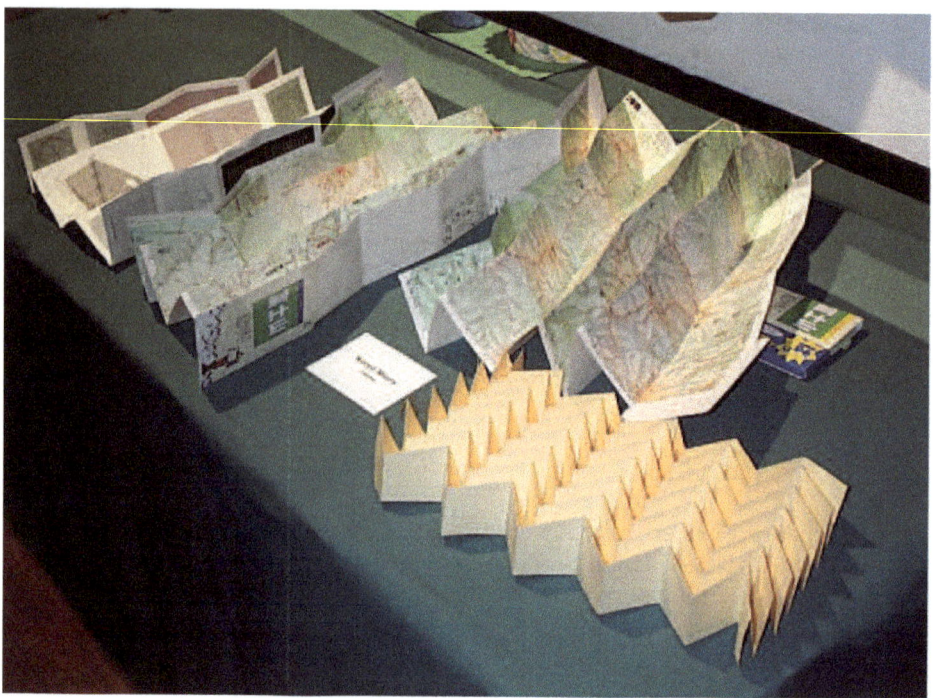

Figure 5.9 A sample of maps folded by the Miura map pattern. [Reprinted by permission of the photographer, Erik Demaine.]

If cardstock paper is creased sharply and cleanly according to the Miura pattern and squeezed between diagonally opposite corners, it collapses beautifully while retaining face rigidity, as depicted in Figure 5.10.

A 3×2 Miura crease pattern is shown in Figure 5.11. The pattern is composed of repeating units each centered on a degree-4 vertex where four congruent rhombs[1] meet, forming a double chevron shape. The vertex is called a ***bird's foot vertex*** in the origami literature because the creases resemble a bird's foot imprint: See ahead to Figure 5.16. We'll call each unit of four congruent rhombs a ***Miura-unit***.[2] One unit is shaded in Figure 5.11.

There is not just one, unique Miura-unit pattern, but rather a continuum of patterns based on the angle θ in the corner of one rhomb, which determines the shape of each rhomb, and so determines the entire Miura-unit shape. In

---

[1] A rhombus (plural rhombs or rhombi) is a parallelogram with equal-length sides.
[2] Miura himself viewed a unit as the feathers of an arrow.

5.3. Miura Map Fold

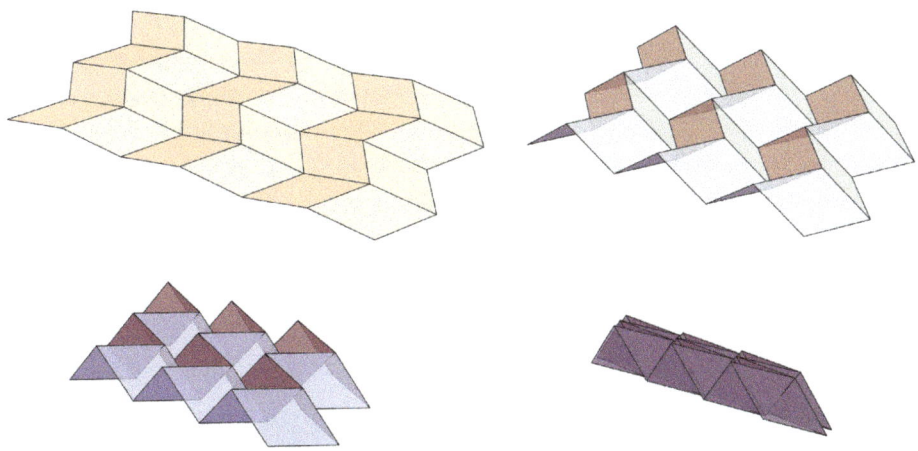

Figure 5.10 Collapsing of a 3 × 2 Miura map fold. Animation: https://cs.smith.edu/~jorourke/MathOrigami/.

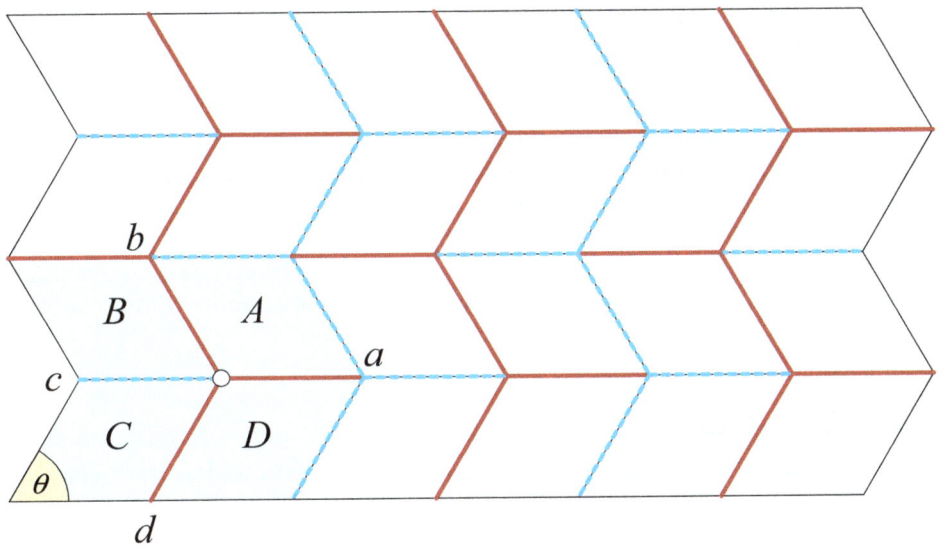

Figure 5.11 Muira 3 × 2 Template. Major creases run vertically, alternately all M (red) and all V (blue). Minor creases run horizontally. Here $\theta = 60°$.

Figure 5.11, $\theta = 60°$, while in Figure 5.12, $\theta = 85°$. We will mainly focus on the case $\theta = 60°$, which enjoys special properties.

Our goal in this section is to analyze the dynamics of the Miura pattern, in particular, to understand why it has 1-DOF, and how the dihedral angles at the creases relate to one another during folding and unfolding.

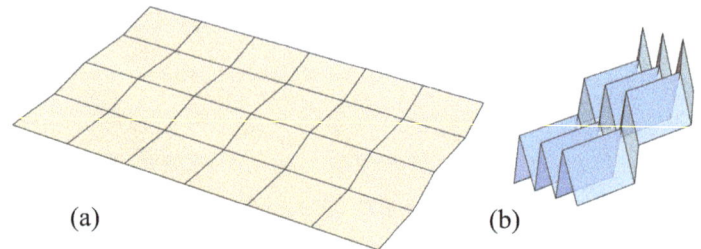

Figure 5.12 Miura pattern with $\theta = 85°$. (a) Initial configuration. (b) Approaching the flat state.

> **Exercise 5.2 [Understanding] Miura Area**
>
> For $\theta = 60°$, calculate the area of the Miura-ori pattern opened flat, before folding (Figure 5.11), and when completely folded flat again (Figure 5.10). Assume all edges have length 1, and the pattern consists of m (nonoverlapping) Miura-units horizontally and n Miura-units vertically.

Each Miura-unit of the Miura pattern is a particular instance of a degree-4 vertex. Because the pattern folds flat, it must satisfy Maekawa's and Kawasaki's Theorems: $M = 3$ and $V = 1$ satisfy Maekawa, and walking around the vertex leads to $60° + 120° = 60° + 120°$ satisfying Kawasaki. But these two theorems do not determine the intermediate dynamics. The sometimes intricate dynamics of rigid origami constructions will be the focus of the remainder of this chapter.

5.3.1 Plus-Sign

First we study the very special Miura-unit when $\theta = 90°$, and so the creases at the vertex are each separated by $90°$. We call this the **plus-sign** unit. Our goal in this section is to understand the 3D configurations achievable by folding the creases, a warm-up preparation for analyzing Miura-units for any θ.

We label the crease endpoints a,b,c,d and v the vertex at the origin, as in Figure 5.13. Take all the edges unit length: $|va| = |vb| = |vc| = |vd| = 1$, and let $\alpha, \beta, \gamma, \delta$ be the dihedral angles along the four edges to a,b,c,d respectively. All dihedrals are flat $180°$ before folding.

Either experimenting with paper or just mentally imagining the constraints leads to the natural hypothesis that segments va and vc fold to the same dihedral angle $\alpha = \gamma$, while vb and vd remain flat with dihedral $180°$. (Of course the same holds for creasing (b,v,d) instead, by symmetry.) And nothing else is possible: Folding va and vc differently is impossible, as is folding vb or vd nonflat.

5.3. Miura Map Fold

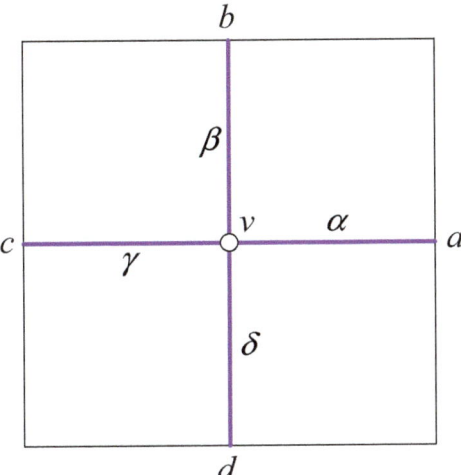

Figure 5.13 Plus-sign; v is the degree-4 vertex at the origin.

We state these intuitive relationships in a formal lemma. Define a dihedral angle α as **extreme** if it is either 180° (flat) or 0° (fully folded), and **nonextreme** or intermediate if $0° < \alpha < 180°$.

> **Lemma 5.1 Plus-Sign Dihedrals**
>
> In the plus-sign pattern, if α is nonextreme, then opposite dihedral angles are equal, and only one pair is nonextreme: More specifically,
>
> (a) $\gamma = \alpha$, and
>
> (b) $\beta = \delta$ is either 180° or 0°.

The lemma states the general situation. We leave the two boundary cases, when $\alpha = 180°$ or when $\alpha = 0°$, to Exercise 5.3.

Now we prove Lemma 5.1. We first argue that points $\{a,v,c\}$ are collinear. Because α is neither 180° nor 0° by assumption, points $\{b,v,d\}$ are not collinear, and so determine a plane P. The two segments va and vb form a right angle at v. So if A rotates about axis va, vb sweeps out the same plane P, as illustrated in Figure 5.14(a). This shows that va is perpendicular to P. Similar reasoning on the other, $B \cup C$ side of P, concludes that vc is perpendicular to P. So then $\{a,v,c\}$ are collinear.

Now suppose, in contradiction to claim (a) of Lemma 5.1, that $\gamma \neq \alpha$. Then the two squares B and C cannot align and mate with the A and D squares along the (b,v,d) path, as is evident from Figure 5.14(b), because of the angle mismatch. So $\gamma = \alpha$, and claim (a) is settled.

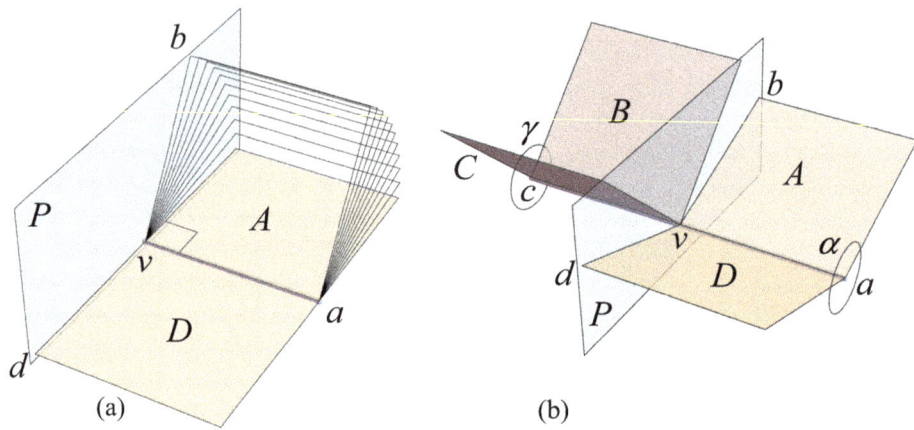

Figure 5.14 (a) Segment va is perpendicular to plane P. (b) $\alpha = 160°$ and $\gamma = 135°$ creates a mismatch along the path (b,v,d).

Turning to claim (b), suppose β is nonextreme, contradicting the claim. Then the positive fold along vb makes $\{a,v,c\}$ noncollinear, contradicting our earlier conclusion. The same contradiction is achieved if δ is nonextreme So it must be that $\beta = \delta$ is extreme and (b) is established. Lemma 5.1 is proved.

We offer a more succinct version of the proof using different terminology in Box 5.2.

> **Exercise 5.3 [Practice] Plus-Sign Proof**
>
> Describe the configuration of the plus-sign unit when $\alpha = 180°$ (i.e., va is uncreased), and when $\alpha = 0°$ (i.e., va is fully folded).

> **Box 5.2 Plus-Sign Lemma**
>
> Let v_1, v_2, v_3, v_4 be the vectors from the degree-4 vertex to a,b,c,d respectively. If the angle α along v_1 is nonextreme, then v_2 and v_4 are not parallel, and so define a plane P as shown in Figure 5.15. Because v_1 is perpendicular to both v_2 and v_4, v_1 is perpendicular to P. Similarly, v_3 is perpendicular to P. Therefore v_1 and v_3 are parallel, which implies that v_2's angle β is extreme.
>
> The remainder follows the argument in the text.

We will see that the opposite dihedral angles have the same magnitude in the more general Miura-unit, when $\theta \neq 90°$.

5.3. Miura Map Fold

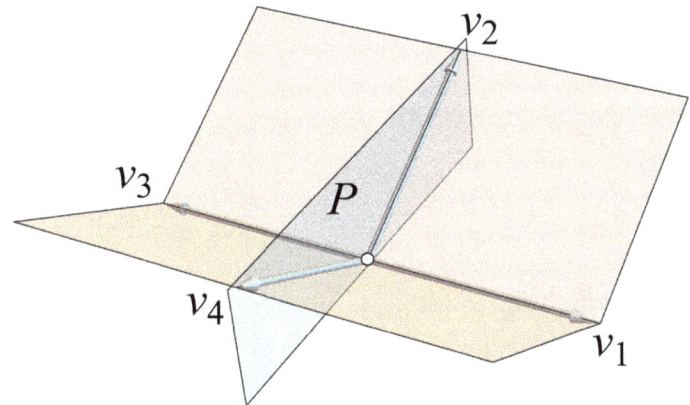

Figure 5.15 Alternative concise proof of Lemma 5.1.

5.3.2 Dynamics of a Miura-Unit

Now we turn to analyzing the dynamics of an isolated, single Miura-unit, which we'll see later (Section 5.3.6) is the key to the behavior of a Miura map fold composed of many units.

We again fix θ to 60° so the four unit-length rhombs have two angles 60° and two angles 120° incident to the degree-4 vertex at the origin v on the xy-plane, with the z-axis upward. We'll use the angle ϕ representing the rotation of edge va downward, with symmetric rhombs A and D at the same z-depth as a, and forming the dihedral α. This will serve as our 1-DOF, ϕ controlling α. We fix b and d to remain on the xy-plane throughout the motion. See Figure 5.16.

Our goal now is to describe these three claims:

> **Theorem 5.1 Miura-Unit Dynamics**
>
> Folding a Miura-unit satisfies these properties:
>
> (1) Opposite dihedral angles are equal in magnitude.
>
> (2) A Miura-unit evolves as a 1-DOF motion.
>
> (3) The relationship between the dihedrals follows a half-tangent equation.

We will prove (1) and (2), and discuss but not prove (3).

5.3.3 (1) Equal Dihedrals

Note that the plus-sign proof (Lemma 5.1) used the collinearity of $\{a,v,c\}$ as a first step, which does not hold when $\theta \neq 90°$. Nevertheless, that proof provides a model for the second step, mating the (b,v,d) paths.

72   5 *Rigid Origami and Degree-4 Vertices*

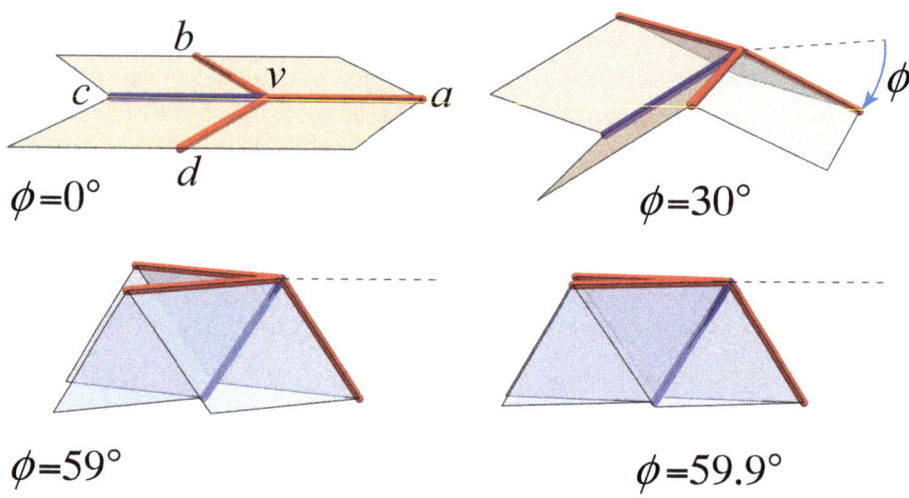

Figure 5.16 Four positions of one Miura-unit, determined by ϕ (which determines α, the dihedral along va). Animation: https://cs.smith.edu/~jorourke/MathOrigami/.

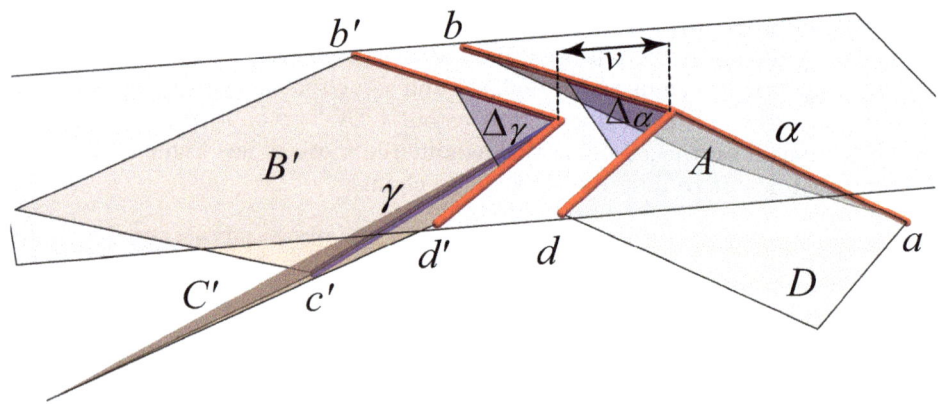

Figure 5.17 Separating $B \cup C$ from $D \cup A$ along path (b,v,d). Corresponds to $\phi = 30°$ in Figure 5.16.

For a positive dihedral angle $\alpha < 180°$, let Δ_α be the angle at v of the triangle dvb. In our setup, this triangle lies in the xy-plane; see Figure 5.17. This angle Δ_α is not identical to α but is uniquely determined by α.[3] The path (b,v,d) forms a boundary component of the $D \cup A$ rhombs.

---

[3] See Box 5.3.

5.3. Miura Map Fold

Now imagine disconnecting the Miura-unit along this path and separating $B \cup C$ from $D \cup A$, as in Figure 5.17. Label the moved points and rhombs with primes. In order for the shifted path (b',v',d') to rejoin with (b,v,d), it must be that the angle Δ_γ at v' is equal to Δ_α. Now Δ_γ uniquely determines γ, so we have symbolically:

$$\alpha \longleftrightarrow \Delta_\alpha = \Delta_\gamma \longleftrightarrow \gamma .$$

Thus $\alpha = \gamma$.

The same logic can be applied to prove that $\beta = \delta$. The triangle avc lies in a vertical plane separating $A \cup B$ from $C \cup D$. The angle Δ_β of triangle avc at v must match the corresponding angle Δ_δ for the two halves to join along the path (a,v,c). Because Δ_β and Δ_δ are each uniquely determined by β and δ respectively, we have

$$\beta \longleftrightarrow \Delta_\beta = \Delta_\delta \longleftrightarrow \delta .$$

Thus $\beta = \delta$ and claim (1) of Theorem 5.1 is settled.

Box 5.3 One-to-One Function

The phrase "α uniquely determines Δ_α" means that there is a **one-to-one function** between α and Δ_α: for each α there is a unique Δ_a, and for each Δ_α there is a unique α. See Box 5.4 for the explicit function.

Box 5.4 α Determines Δ_α

A bit of messy trigonometry eventually yields this simple relationship between α and Δ_α:

$$\Delta_\alpha = \arccos\left(\frac{1}{4}(3\cos(\alpha) + 1)\right) .$$

Figure 5.18 shows that indeed α uniquely determines Δ_α. And the reverse holds: Δ_α uniquely determines α. As a check, note when $\alpha = 180°$, $\Delta_\alpha = 120°$, which is the angle at v between the two rhombs D and A when the Miura-unit is opened flat. And when $\alpha = 0°$, so is $\Delta_\alpha = 0°$, corresponding to $\phi = 60°$ in Figure 5.16.

5.3.4 (2) 1-DOF

Knowing that opposing dihedrals are equal in magnitude still leaves the possibility that α and β (and therefore γ and δ) are independently adjustable, i.e., that the motion has two degrees-of-freedom. But now we argue that a Miura-unit (for any θ) has just one degree-of-freedom, 1-DOF.

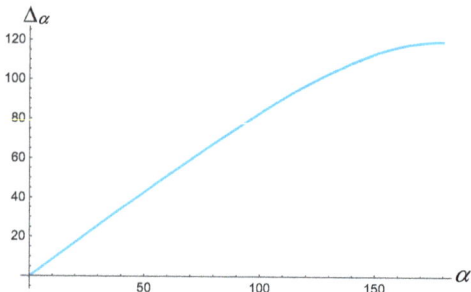

Figure 5.18 α uniquely determines Δ_α.

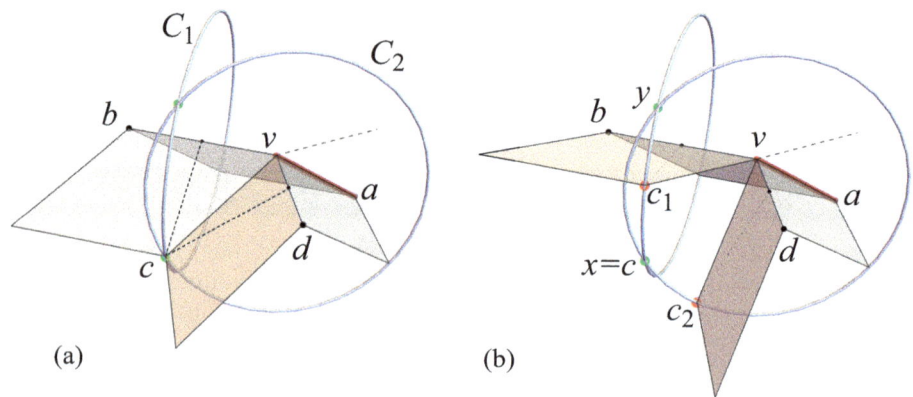

Figure 5.19 The two circle intersections at x and y represent the possible positions of c, with position y violating the V-fold along vc.

Fix α which fixes the positions of A and D. Detach rhombs B and C from one another by slicing their common edge vc, replicating point c to two copies c_1 and c_2. Let B and C now swivel independently around edges vb and vd respectively. These hinge motions have c_1 and c_2 rotating on vertical circles C_1 and C_2 centered on the midpoints of vb and vd respectively, as shown in Figure 5.19(b). This swiveling contradicts the 1-DOF hypothesis: α is already one degree-of-freedom, and the independent swivel would be a second degree-of-freedom. So we aim to show that $c_1 = c_2$, i.e., there is no swiveling possible.

The reason the C_1 circle is centered on the midpoint of vb is that the $\sqrt{3}/2$ altitude of triangle vbc (half of rhomb B) is perpendicular to vb at its foot, the midpoint. See Figure 5.19(a). So swiveling B around axis vb traces out that C_1 circle. Similar claims hold for the C_2 circle.

The two circles intersect in two points, labeled x and y in Figure 5.19, where x is the predetachment position of c. This follows from intersecting the two vertical planes containing C_1 and C_2: They intersect in a vertical line that passes through x and y.

5.3. Miura Map Fold

The only way to reattach the separated B and C rhombs is to have c_1 and c_2 coincide at $c = x$. To have $c = y$ violates the stipulation that vc is a V-fold. Thus indeed the configuration of the entire unit is determined by the 1-DOF α. This settles (2) of Theorem 5.1.

> **Exercise 5.4 [Understanding] MM Configuration**
>
> Describe the shape of the Miura-unit when $c = y$ in Figure 5.19.

5.3.5 (3) Half-Tangent Equation

What remains is claim (3) of Theorem 5.1: The relationship between the dihedrals follows a half-tangent equation. Specifically there is a remarkable relationship between dihedrals α and β. We know that α determines β from the 1-DOF result. But naively there is no reason to expect a simple relationship. In a $\theta = 60°$ Miura-unit, the following holds throughout the motion:

$$\tan(\alpha/2) = 2\tan(\beta/2) . \tag{5.1}$$

One consequence is that $\alpha > \beta$ in any configuration of the Miura-unit: see Box 5.5.

> **Box 5.5 Half-Tangent Equation**
>
> Both α and β range from $0°$ (folded flat) to $180°$ (uncreased), so the half-angles range from $0°$ to $90°$. The tangent function is strictly increasing in this range. Because $\tan(\alpha/2)/\tan(\beta/2) = 2$, it must be that $\tan(\alpha/2) > \tan(\beta/2)$. Because the function is increasing, this implies that $\alpha/2 > \beta/2$, so $\alpha > \beta$.
>
> Because these are dihedral angles, the va α-crease is flatter than the vb β-crease, or equivalently, the vb crease is folded more sharply than is the va crease.

Here are two sample evaluations, the first when α is still large (recall initially $\alpha = 180°$), and the second when α is approaching $0°$.

(1) When $\alpha = 141°$, then $\beta = 109°$ and

$$\tan(141°/2) \approx 2\tan(109°/2)$$
$$2.828 \approx 2 \cdot 1.414.$$

(2) When $\alpha = 32°$, then $\beta = 16°$ and

$$\tan(32°/2) \approx 2\tan(16°/2)$$
$$0.287 \approx 2 \cdot 0.143.$$

Although I know of no simple proof of this beautiful tangent relationship, it can at least been seen as plausible by the following reasoning, supported by this second example where the collapsing to planarity is nearly complete. First, the tangent of a small angle ε is nearly ε. So for small angles, Equation (5.1) says that $\alpha = 2\beta$. Now look at the $\phi = 59°$ folding in Figure 5.16. Here it is almost evident that the dihedral α is composed of β and δ together: $\beta + \delta = 2\beta$.

Equation (5.1) generalizes to Miura-units with any particular θ:

$$\tan(\alpha/2) = \mu \tan(\beta/2)$$

for some constant **multiplier** μ. We see $\mu = 2$ because of our focus on $\theta = 60°$. For $\theta = 85°$ (Figure 5.12), $\mu \approx 11.47$. The multiplier μ depends only on θ.

In fact it is remarkable that Theorem 5.1 and the half-tangent equation generalize to any degree-4 vertex v that is flat-foldable (and so satisfies Kawasaki's Theorem and the other results established in Chapter 3). Then the multiplier μ is determined by the four **sector** angles incident to the vertex v. Once those sector/face angles are fixed, then the entire dynamics of the vertex's motion is determined as in Theorem 5.1. We will invoke this generalized version as Theorem 5.2 in Section 5.4.2.

> **Exercise 5.5 [Understanding]** $\theta \to 90°$
>
> Comparing $\theta = 60°$ (Figure 5.10) with $\theta = 85°$ (Figure 5.12), what happens as θ gets closer and closer to $90°$, in symbols, $\theta \to 90°$?

5.3.6 Tessellations of Miura-Units

We return to the 3×2 tessellations of Miura-units earlier displayed folding flat in Figures 5.10 ($\theta = 60°$) and 5.12 ($\theta = 85°$). We claimed in Section 5.3 that a tessellation of repeated Miura-units has just one degree-of-freedom regardless of the number of repeated units. We are now in a position to justify that claim.

Recall the Miura-unit in the corner of the template in Figure 5.11, repeated with additional details in Figure 5.20. Call that corner unit of four rhombs M_o. Let o be the vertex of M_o, and o_x and o_y the vertices one unit to the right and up respectively. Then rotating a copy M_x of M_o counterclockwise about the midpoint of oo_x places the α dihedral angle of M_x on top of the α dihedral of M_o. Now changing α controls the configurations of both M_o and M_x by the 1-DOF result of Theorem 5.1(2).

Similarly, rotating a copy M_y counterclockwise about the midpoint of oo_y places the β dihedral angle of M_y on top of the β dihedral of M_o. So changing β of M_o makes the same changes to β of M_y. In this manner, the 1-DOF each unit overlaying matching dihedral angles. And this is why tugging on opposite corners collapses and reopens the entire tessellation in a pleasing coordinated motion. Again see the animation snapshots in Figures 5.10 and 5.12. We will

5.4. *The Square Twist* 77

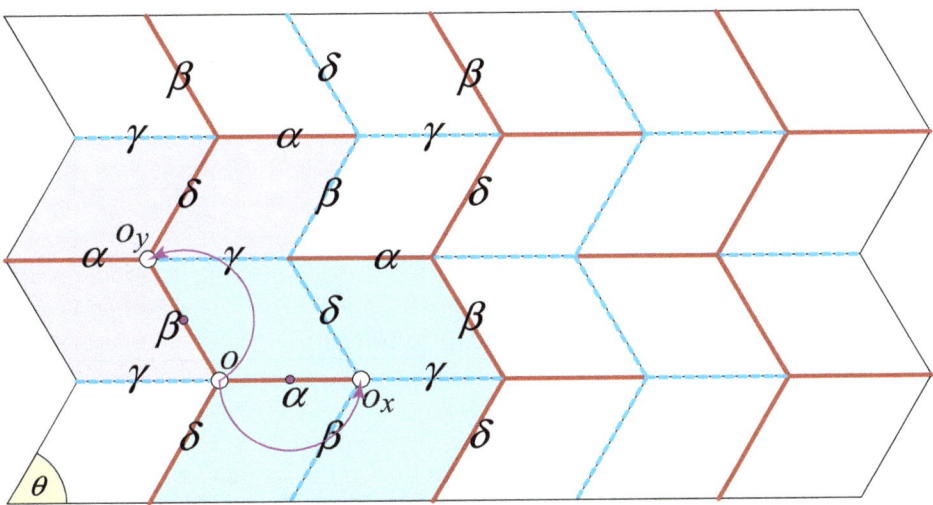

Figure 5.20 Rotating Miura-unit M_o in corner overlays M_x (teal) and M_y (purple) matching dihedrals α and β respectively.

see in Chapter 9 that pragmatically, the coordinated motion is easier to achieve opening a collapsed tesselation than closing a fully opened-flat configuration.

In a Miura tesselation, the vertical dihedrals β, δ are all M-folds or all V-folds. These are called ***major*** creases, while the horizontal α, γ folds alternate M and V. These are called ***minor*** creases. Note that, because $\alpha > \beta$ (Equation (5.1)) and α and β are dihedral angles, β is folded more sharply than is α, justifying calling it the "major" crease.

5.4 The Square Twist

The next example of rigid origami we'll explore is the ***square twist***, which exhibits an attractive twisting motion. Like the Miura example, it has a single degree-of-freedom and can be repeated to a tessellation. Unlike the Mirua-unit, a square-twist unit involves four degree-4 vertices.

Figure 5.21(a,b) shows two different M/V assignments to the same underlying crease pattern. The pattern consists of four degree-4 vertices, each with the same incident sector angles: $90°, 45°, 90°, 135°$. M/V assignment (a) is called the M^4 pattern, reflecting the four M assignments around the square, while (b) is called the M^2V^2 pattern. Despite the similarity between (a) and (b) of the figure, the patterns behave rather differently.

First we note that the two M/V assignments both satisfy Maekawa's and Kawasaki's Theorems and the Local-Min Lemma from Chapter 3, so both can fold flat: Exercise 5.6. Figure 5.21(c,d) shows the two flattenings. The more symmetric M/V assignment in Figure 5.21(a) folds with the central square diamond on top, whereas the M^2V^2 assignment folds so that only one-quarter of the central diamond is exposed from above.

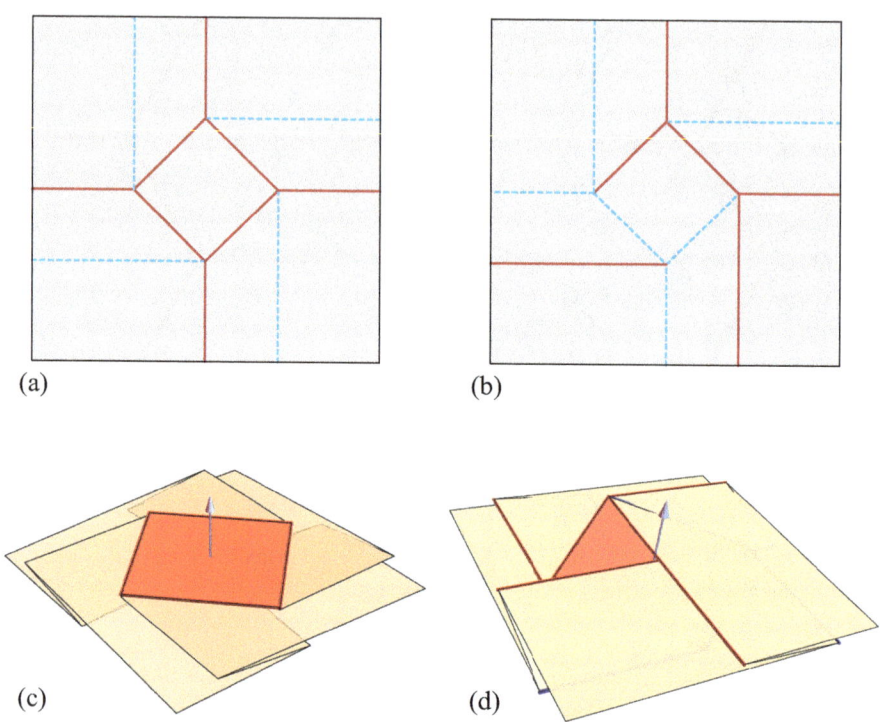

Figure 5.21 Square twist templates. (a) M^4 pattern (not rigidly foldable). (b) M^2V^2 pattern (rigidly foldable). (c,d) Flat foldings of (a,b) respectively. (Vectors just to add 3D perspective.)

As with the Miura-unit, a tug on opposite corners of the M^4 pattern in Figure 5.21(c) twists the square 90° in a pleasing 1-DOF motion, until all dihedral angles are opened to 180° as in Figure 5.21(a). However, if you make this model out of paper, one can feel a slight hitch in this motion. In fact, this action *cannot* be achieved with rigid faces, for a subtle reason that we'll discuss later (Section 5.4.3). Opening and closing the model bends the central square along a diagonal, and if the square were a rigid plate, the motion would bind. Surprisingly, if one adds a horizontal crease to the diamond in Figure 5.21(a), so that the center is now two hinged right triangles, with degree-5 vertices at the endpoints of the shared crease, it then is rigidly foldable. Henceforth we concentrate on the M^2V^2 pattern in Figure 5.21(b).

> **Exercise 5.6 [Practice] Flat Folding Degree-4 Vertex**
>
> Show that each degree-4 vertex in both Figure 5.21(a) and (b) satisfies Maekawa's Theorem 3.1, Kawasaki's Theorem 3.2, and the Local-Min Lemma 3.2.

5.4. The Square Twist

> **Exercise 5.7 [Understanding] Square Twist Area**
>
> Calculate the area reduction from the unfolded pattern in Figure 5.21(a), to the fully folded state in Figure 5.21(c). Assume the central square has unit-length edges, and that the unfolded square paper has side length $2 + \sqrt{2}$.

5.4.1 Rigidly Foldable Square Twist

The M^2V^2 M/V assignment, shown in Figure 5.21(b), *is* rigidly foldable. Two of the central square's edges are V-folds, and so the square does not sit on top in the final folded state, as we saw in Figure 5.21(d).

However, because the pattern is rigidly foldable, the action initiated by tugging on opposite corners produces a smooth opening motion, without the hitch experienced with the non-rigid M^4 pattern. The reader is encouraged to fold the template in Figure 5.21(b) or view the animation ahead in Figure 5.23.

Next we analyze the dynamics of one degree-4 vertex of the rigid square twist. Note that two of the four degree-4 vertices in the M^2V^2 pattern (Figure 5.21(b)) have $3M + 1V$ incident folds, and two have $3V + 1M$ incident folds. The dynamics surrounding the four vertices is the same, except two are the flip side of the other two. We call the four faces of a 3M vertex v a ***square-twist unit***.

5.4.2 Dynamics of a Square-Twist Unit

We mentioned that the dynamics of folding a Miura-unit described in Theorem 5.1 hold more generally for any degree-4-unit that is flat-foldable. Because we know the M^2V^2 pattern is flat-foldable (Exercise 5.6), this generalized theorem holds. We will not prove this claim, only state it.

> **Theorem 5.2 Degree-4 Folding**
>
> Folding a flat-foldable degree-4 vertex satisfies these properties:
>
> (1) Opposite dihedral angles are equal in magnitude: $\alpha = -\gamma$ and $\beta = \delta$.
>
> (2) The configuration evolves as a 1-DOF motion.
>
> (3) The relationship between the dihedrals follows the half-tangent equation.
>
> $$\tan(\alpha/2) = \mu \tan(\beta/2),$$
>
> where the multiplier μ depends solely on the face angles incident to v.

Knowing that opposite dihedrals are equal in magnitude, and knowing the half-tangent equation holds, permits computing the dynamics of all four dihedrals from any one of them.

The multiplier μ in the half-tangent Equation (5.1) for a square-twist unit, is $\sqrt{2}+1$, whereas it is 2 for the Miura-unit:

$$\tan(\alpha/2) = (\sqrt{2}+1)\tan(\beta/2). \tag{5.2}$$

This reflects the fact that the multiplier depends on the sector angles incident to the vertex v, and the sector angles for the Miura-unit are different than those for the square twist.

Two examples:

(1) When $\alpha = 80°$, $\beta \approx 38.33°$:

$$\tan(80°/2) \approx (\sqrt{2}+1)\tan(38.33°/2)$$
$$0.839 \approx 2.414 \cdot 0.348.$$

(2) The second example corresponds to Figure 5.22(b), when $\alpha, \beta = 135°, 90°$:

$$\tan(135°/2) = (\sqrt{2}+1)\tan(90°/2)$$
$$2.414 \approx 2.414 \cdot 1.$$

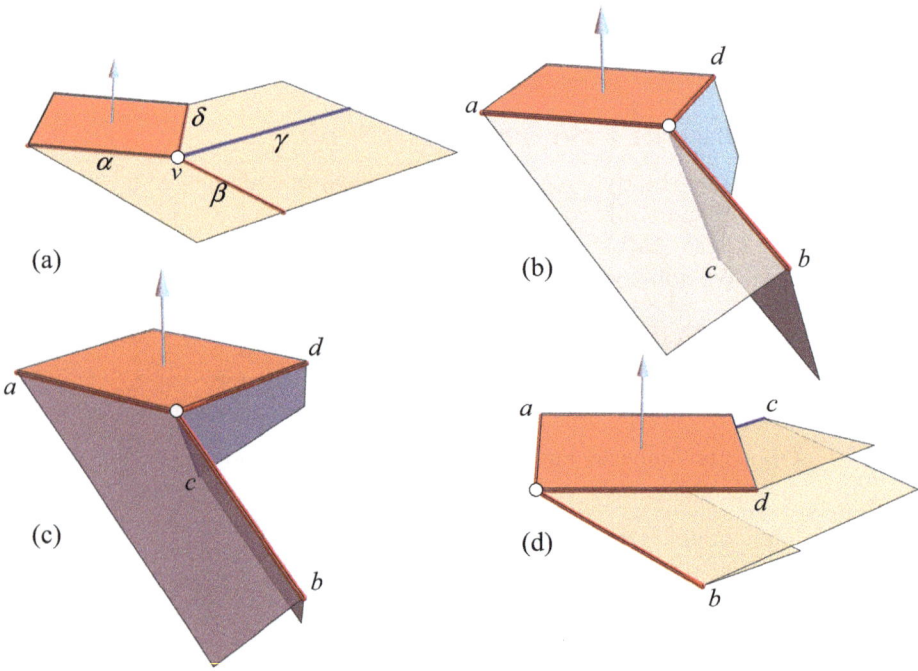

Figure 5.22 Snapshots of a square-twist unit folding: $\alpha = 180°, 135°, 90°, 10°$. Animation: https://cs.smith.edu/~jorourke/MathOrigami/.

5.4. The Square Twist

In both cases, $\alpha > \beta$ (α is flatter than β) for the same reasons as detailed in Box 5.5.

Concerning the Degree-4 Folding Theorem 5.2, master origamist Robert Lang says:

> "I think this result is one of the most beautiful results in all of mathematical origami due to its unexpected simplicity and symmetry." (Lang, 2017, p. 491).

5.4.3 Rigidly Foldable Square Twist

Now we return to the square-twist template in Figure 5.21(b), and put four square-twist units together. In Figure 5.23(a), we label the faces incident to the

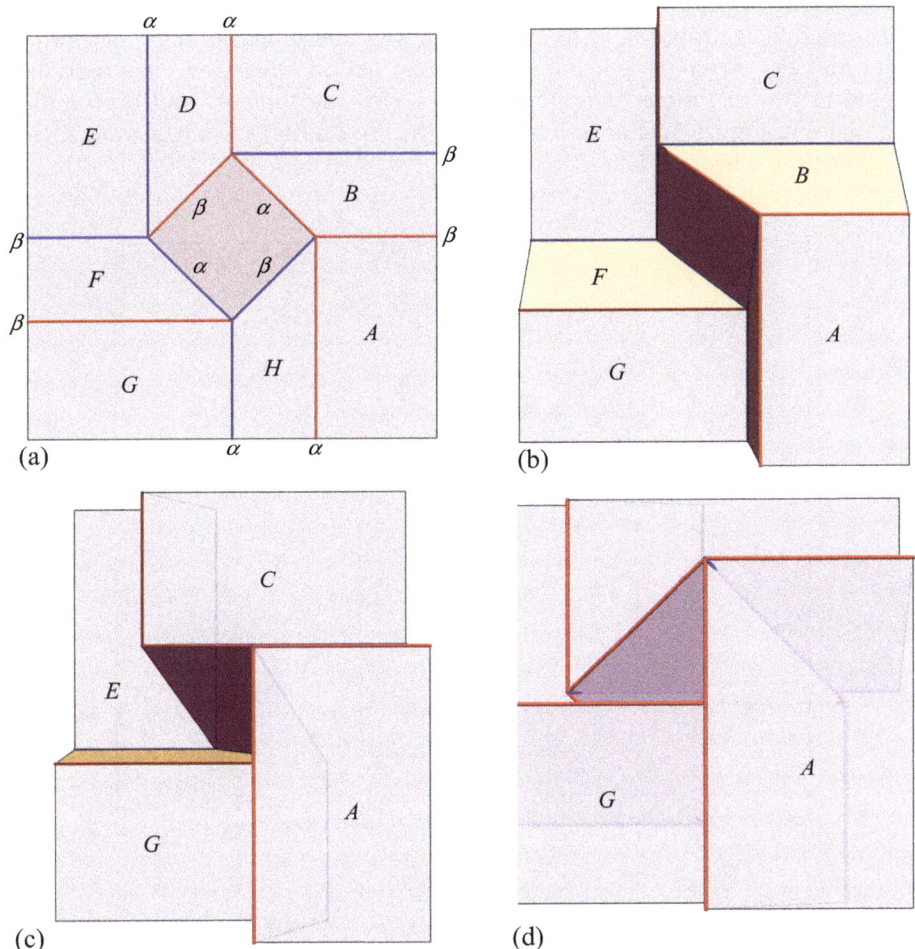

Figure 5.23 Snapshots of rigid folding of a square twist. $\alpha = 180°, 90°, 45°, 10°$. Animation: https://cs.smith.edu/~jorourke/MathOrigami/.

central square, as well as the dihedral angles α, β inherited from each degree-4 vertex. So face A is bounded by two M-folds, but the vertical edge has dihedral α and the horizontal edge has dihedral β. Note the vertical α-folds are all M or all V, while the horizontal β-folds alternate M/V. The snapshots from an animation in the figure show the central square twisting 90° counterclockwise, eventually (in Figure 5.23(d)) being three-quarters obscured by faces A, G, and H behind A.

5.4.4 Square Twist Tessellation

As with the Miura-unit, the square twist can be repeated in an $m \times n$ pattern, tiling a portion of the plane. Figure 5.24 shows a 2×2 tiling. Note that the basic pattern (Figure 5.21(b)) must be reflected so that the M/V and α/β creases join one another consistently. Once this is arranged, the 1-DOF guaranteed by Theorem 5.2(2) propagates, so that fixing any one dihedral angle determines them all. The result is a beautiful dynamic action when the construction is tugged in two directions at opposing spots, say, the topmost and bottommost rectangles in Figure 5.24 or in Figure 5.25(a). You have to feel it in your fingers to appreciate Theorem 5.2.

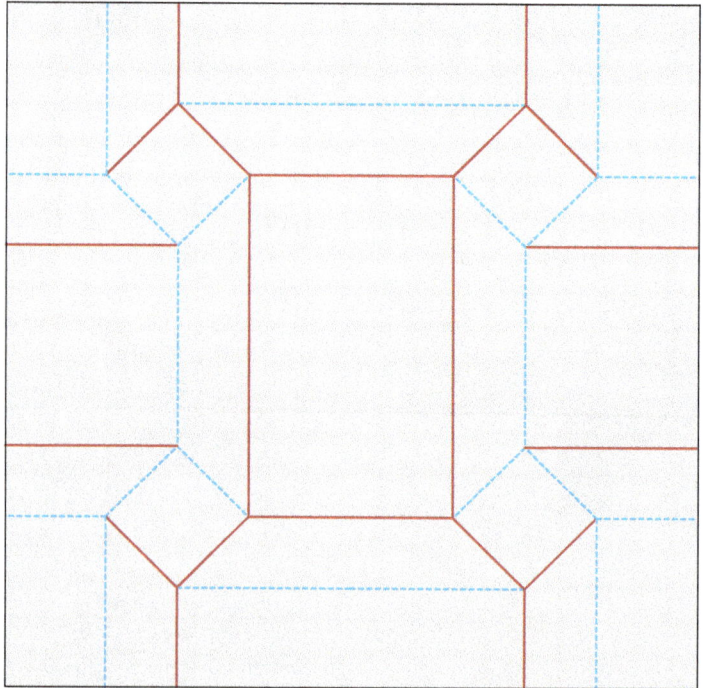

Figure 5.24 A 2×2 tessellation composed of four square twists (Figure 5.21(b)).

5.4. The Square Twist

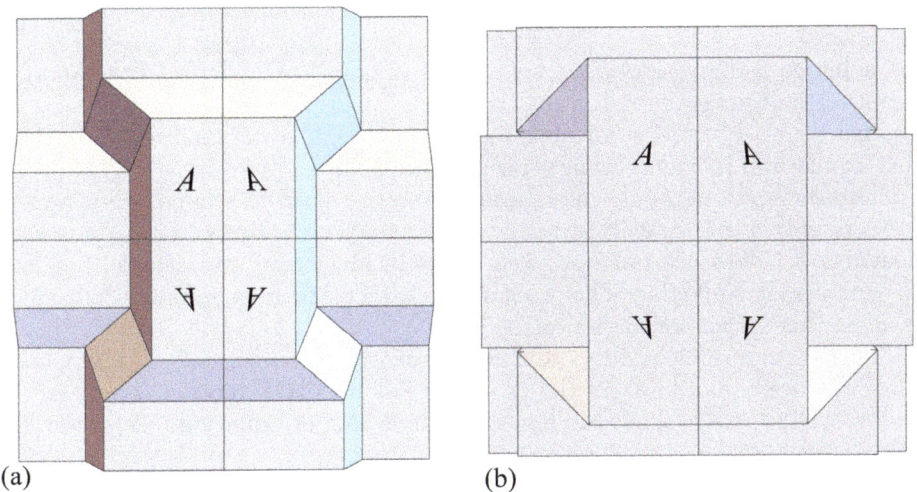

Figure 5.25 (a) $\alpha = 120°$ (left vertical edge of face A), (b) $\alpha = 10°$. Face A in Figure 5.23 is shown in four orientations. Animation: https://cs.smith.edu/~jorourke/MathOrigami/.

We return to the issue of why the M^2V^2 crease assignment in Figure 5.21(b) is rigidly foldable but the M^4 pattern in Figure 5.21(a) is not. We've seen that the four degree-4 vertices each fold flat rigidly in isolation (Figure 5.22). But of course a square twist has four such degree-4 vertices that must follow the same constraints. It turns out that the natural M^4 pattern in Figure 5.21(a) leads to the contradictory relationship $\alpha > \alpha$ when applied to all four vertices, a claim we relegate to a (challenging) Exercise (5.8). However, the M^2V^2 pattern in Figure 5.21(b) of the same figure does not lead to this contradiction.

> **Exercise 5.8 [Challenge] Nonrigid Pattern**
>
> Sketch out an argument for the "race condition" that leads to $\alpha > \alpha$.

No Rigidly Foldable Triangle Twists It is natural to hope for a rigidly foldable triangle twist. We already encountered a triangle twist in the previous chapter: Section 4.5.2 and especially Figure 4.15. Alas, that triangle twist is not rigidly foldable. Moreover, for nonobvious reasons, there is no rigidly foldable triangle twist, regardless of the crease pattern or triangle shape. However, there are rigid twists based on several other four-sided twist centers besides squares: rectangles, parallelograms, isosceles trapezoids, and rhomb twists.

5.5 Octahedron Top

As a bit of a digression from traditional origami, we mention here another beautiful half-tangent relationship that emerges from rigid folding. It is based on the degree-4 vertex v forming the top of a regular octahedron, formed by four equilateral incident triangles; see Figure 5.26.

This differs from the usual origami assumption that 360° of paper surrounds v—here only $4 \cdot 60° = 240°$ of material surrounds v. And note that Maekawa's Theorem 3.1 does not hold: All four edges incident to v are M-folds, but it is nevertheless flat-foldable. The proof of Maekawa's Theorem assumed 360° angle at v, so there is no contradiction.

This structure is rigidly flexible, satisfying Theorem 5.2: (1) Opposite dihedrals are equal, and (2) the structure has a 1-DOF motion. Its dihedrals vary from 180° to 0°. They are related by an analog of Equation (5.2), claim (3) of that theorem:

$$\left(\tan \frac{\alpha}{2}\right)\left(\tan \frac{\beta}{2}\right) = 2.$$

For example, in the figure,

$$\left(\tan \frac{120°}{2}\right)\left(\tan \frac{98.21°}{2}\right) \approx 2$$
$$1.732 \cdot 1.155 \approx 2.$$

Again I know of no simple proof of this relationship.

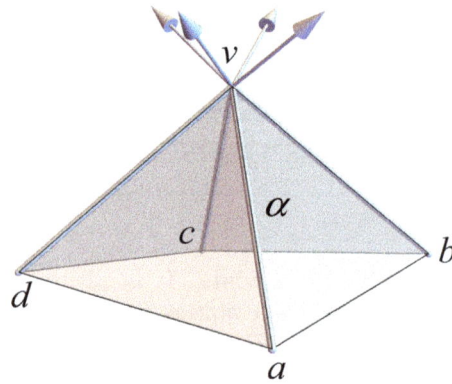

Figure 5.26 Dihedral $\alpha = 120°$. Normal vectors (perpendicular to the four faces) shown. (Recall Box 1.2.) Animation: https://cs.smith.edu/~jorourke/MathOrigami/.

5.6 Rigidly Flat-Foldable Is NP-Hard

As the reader perhaps anticipates from the flat-foldability hardness result in Chapter 4, rigid foldability is also NP-hard. More precisely, it is NP-hard to determine whether a crease pattern composed of degree-4 flat-foldable vertices has an M/V assignment that makes it rigidly foldable. As usual, the proof of this NP-hard result is too complicated to present. But because the proof is a reduction from SET PARTITION, which we already employed in Chapter 4, it seems worthwhile to at least sketch the clever main idea.

Recall the SET PARTITION problem (Section 4.2.1) has for input a set A of positive integers, say, $A = \{1,1,2,3,5,8\}$, and asks if A can be partitioned into two parts with equal sum. In this small case, the answer is easily seen to be YES: $S_1 = \{1,1,3,5\}$ and $S_2 = \{2,8\}$, both of which sum to 10.

Continuing to follow this example, the proof constructs a flat piece of paper with eight internal degree-4 vertices, six for the $n = 6$ integers in A, and two more. See the top left image in Figure 5.27. The sector angles of the six vertices are determined by the integers in A. Each of the S_1 and S_2 partitions is encoded in the pattern by creasing the right sector edge M for each element of S_1 and creasing the left sector edge V for each element not in S_1, and similarly for S_2.[4] See the top row of Figure 5.27.

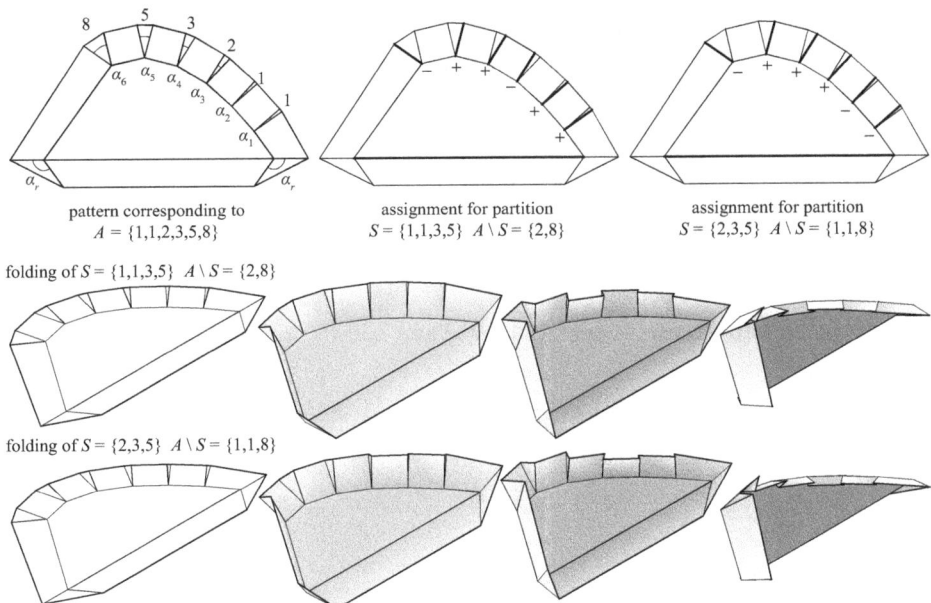

Figure 5.27 Dark segments are M-folds, lighter segments V-folds. [From Akitaya et al. (2020). Reprinted by permission of the authors.]

---

[4] In the figure, the notation $A \setminus S_1$ (A "set minus" S_1) means all the elements of A after removing the elements of S_1.

The authors then prove, using the α, β half-tangent conditions from Theorem 5.2(3), that the construction is rigidly foldable to the flat state if and only if S_1 and S_2 partition A into equal halves. See the bottom two rows of the figure. Note that the multipliers from the half-tangent conditions are all different, because they depend on the sector angles. There are many such details elided here.

Repeating the logic of the reduction from SET PARTITION, we can see that if it were easy to decide by some fast algorithm if a flat-foldable degree-4 crease pattern has an M/V assignment that renders it rigidly foldable, then it would be easy to answer a SET PARTITION instance quickly: Encode the elements of A in the sector angles, and use the fast rigidly foldable algorithm to decide if there is an M/V assignment that partitions A. But SET PARTITION is an NP-hard problem which no one knows how to solve in subexponential time, so there cannot be such a fast algorithm.

The NP-hardness of rigid foldability is yet more evidence that origami is richly complex.

5.7 Technical Notes

Sec. 5.4: The Square Twist Invented independently by Shuzo Fujimoto and by Yoshihide Momotani. Robert Lang's massive book *Twists, Tilings, and Tessellations* (Lang 2017) situates the square twist within a much larger context. The race condition in Exercise 5.8 is from Lang (2017, p. 493).

The useful terminology distinguishing the M^4 and M^2V^2 patterns is due to Lang (2017, p. 492).

Adding a diagonal to the central square to make it rigidly foldable: Hull and Urbanski (2018).

Sec. 5.4.4: No Rigidly Foldable Triangle Twists There are no triangle twists that are rigidly foldable: Lang (2017, p. 510). Other four-sided centers are possible: Larry Howell.

Sec. 5.5: Octahedron Top Proof by Anna Lubiw, personal communication 2023.

Sec. 5.6: Rigidly Flat-Foldable is NP-hard Akitaya et al. (2020).

6
Origami Design

6.1 Introduction

Origami design—the invention of a folding to model a specific target shape—is perhaps more of an art than a science. Or it was until the 1990s, when several origamists developed mathematical techniques that added more science without excluding the art. In this chapter we will explore aspects of the work of famed origamist Robert Lang, who championed a particular approach based on "uniaxial bases," which allowed him to create stunning models such as the deer shown in Figure 6.1. We do not present his most advanced technique, which he calls the "tree method," implemented in an algorithm TREEMAKER. Instead we concentrate on the tools—especially "molecules"—supporting the ***circle/river method*** of design, which includes precursor aspects of the tree method.

We make no attempt to teach how to design origami models, nor even how to fold an already designed model—This is not a how-to book. Rather we follow a particular vein in Lang's work that employs accessible mathematics, and prepares us for the Fold & 1-Cut Theorem in Chapter 7.

One origami mystery will be resolved in this chapter. Naively circles serve no particular function in origami, but we'll gradually see that in fact "circle packing" plays a central role in origami.

6.2 The Design Process

Typically a designer starts (I) by folding the paper square into one of the traditional ***bases*** that somehow fits the target subject, and then (II) folds the details, relying on experience. Each of the half-a-dozen or so traditional bases have relatively simple crease patterns, leaving several "flaps" of paper for manipulation in part (II). The bases have names reflecting simple target shapes—kite, fish, bird, frog, windmill. Figure 6.2 shows two bases. Although the windmill base resembles a windmill, it is not evident that the fish base is close to a fish.

Figure 6.1 Robert Lang: *White-tailed Deer.* Modern version of 1996 original. [Reprinted by permission of Robert Lang.]

The design paradigm we present follows the same two-step process, except now the first step (I) generalizes beyond the traditional bases to bases that match the designer's choice of an appropriate "stick figure." Call this step (I-a). We'll concentrate on constructing a "uniaxial base" compatible with that stick figure, step (I-b), for it is here that we encounter rich mathematics and algorithms.

There remains designer choices in step (I-b), and step (II) is unaltered, relying on the designer's origami experience. So the art in origami design remains.

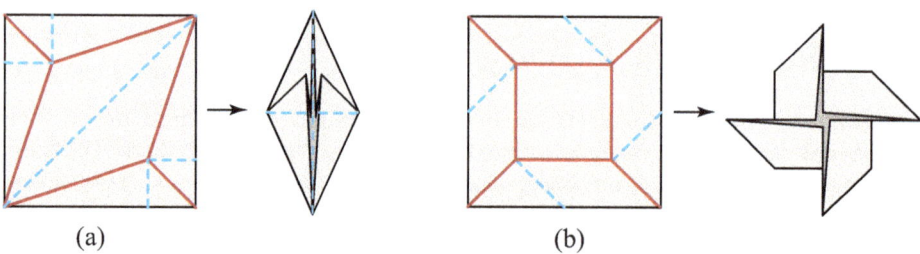

Figure 6.2 (a) Fish base. (b) Windmill base.

6.3. Uniaxial Bases

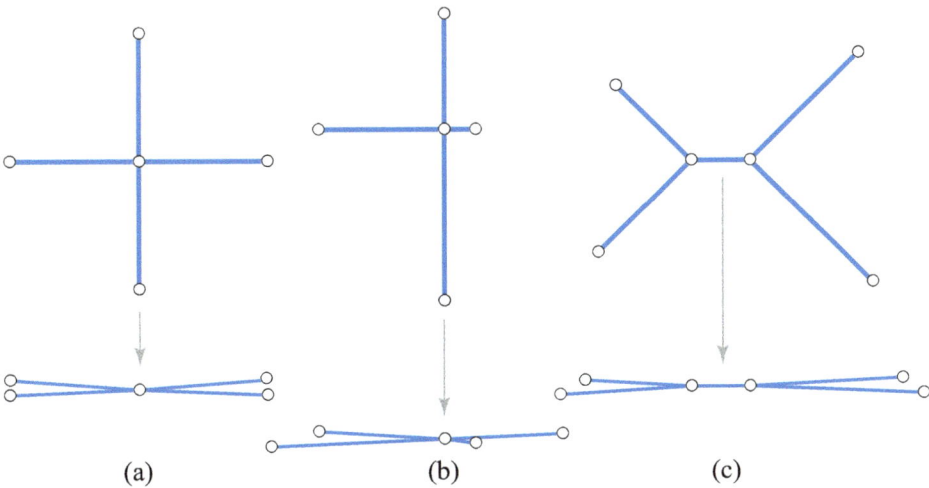

Figure 6.3 Stick figures collapsed to become uniaxial, each on a line.

6.2.1 Stick Figures

A *stick figure* is technically an *embedded metric tree*. We illustrate the meaning of these terms referring to Figure 6.3. For our purposes, a *graph* is a network formed of straight segment edges connected at vertices. A *tree* is a graph that has no cycle, i.e., no closed loop of edges. Trees have *leaves*, nodes with only one incident edge. In a *metric tree*, each edge has a length. Finally, an *embedded tree* lies in the plane with the order of edges about each vertex fixed, but without specifying the angle between adjacent edges. With angles unspecified, a tree can be "collapsed" to lie along a line, as illustrated in Figure 6.3. We'll continue to favor the origami term "stick figure" instead of using graph terminology.

Returning to step (I-a) of the design process, the designer specifies a stick figure, including edge lengths and the ordering of edges around each vertex. In some sense, the stick figure represents the number and lengths of "appendages" of the target model. Step (I-b) creates a *uniaxial base* that projects to the collapsed stick figure in a way we will explain in the next section.

> **Exercise 6.1 [Understanding] Collapsing to a Line**
>
> Why can every tree be collapsed to a line, as illustrated in Figure 6.3?

6.3 Uniaxial Bases

We illustrate design step (I-b) with one of the traditional origami bases, the waterbomb base. A *waterbomb* is a folding of a square to a cube with a hole

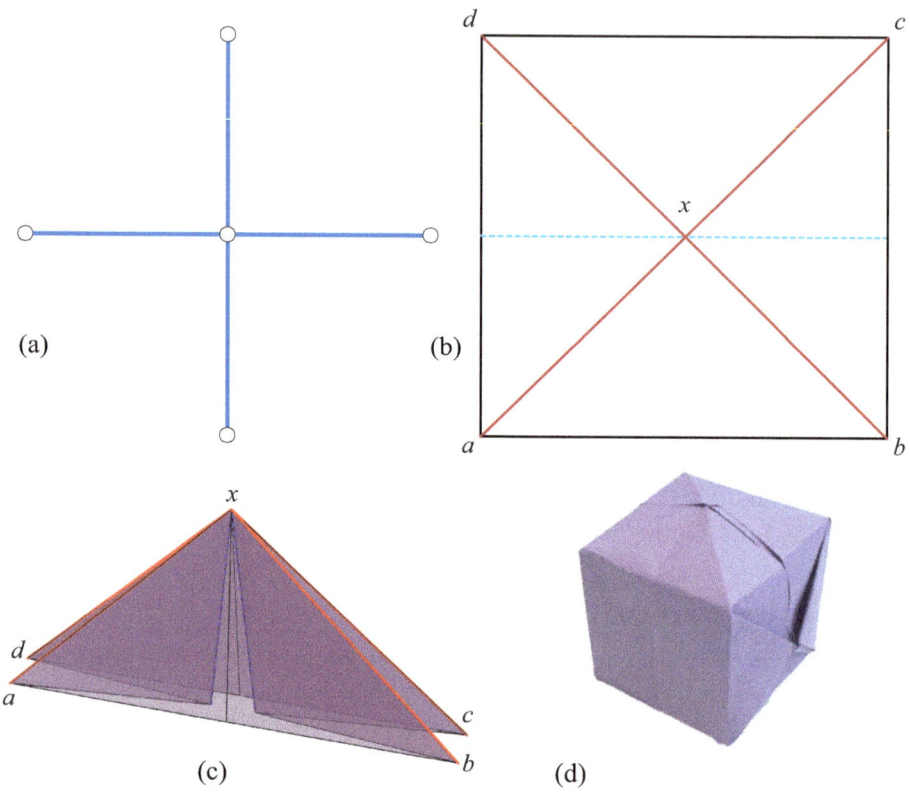

Figure 6.4 (a) Waterbomb stick figure. (b) Base crease patterns. (c) Folding of base. (d) Completed waterbomb.

in the middle of one face; see Figure 6.4(d). The cube can be filled with water through the hole and used in jest as a water balloon or "bomb." The design is more than a century old.

The stick figure for this base is a simple degree-4 cross with equal length edges, shown in Figure 6.4(a) (and Figure 6.3(a)). This leads to a crease pattern for the base, Figure 6.4(b). It is the passage from stick figure (Figure 6.4(a)) to crease pattern (Figure 6.4(b)) where mathematics enters, as we'll explain in Section 6.3.1. When folded, the base satisfies a collection of special and useful *uniaxial base* properties, which we now enumerate.

(1) When the base is collapsed flat, all its flaps lie along a single "axis" line in the xy-plane.

(2) The entire perimeter of the paper lies along this axis.

(3) All the base's valley **hinge creases** are perpendicular to the xy-plane.

6.3. Uniaxial Bases

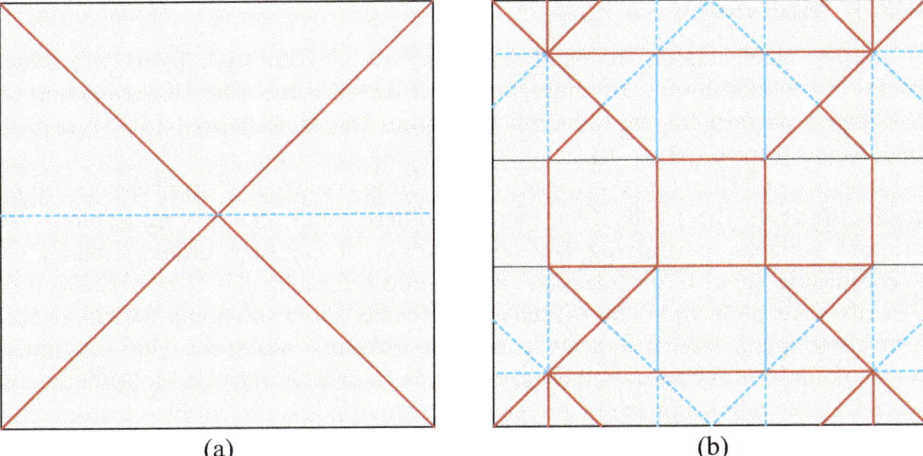

Figure 6.5 (a) Waterbomb base crease pattern. (b) Waterbomb crease pattern.

(4) The shadow of the folding on the plane matches the collapsed stick figure.

(5) The circle tangency points must align with the hinge creases.[1]

In Figure 6.4(c), note in particular that the flaps swivel on z-vertical hinge creases. Now art reenters in step (II), folding the waterbomb from the base, for it is far from obvious how to fold the final model from the base. In the waterbomb, each flap becomes a type of lock covering half of one cube face. So the four flaps together cover two whole faces, leaving four faces covered without creases,[2] evident in the central horizontal strip of the crease pattern in Figure 6.5(b). Perhaps it is easier to see how a pair of connected waterbomb bases are used to create the jumping frog (an elementary origami model), where four of the available flaps become the four legs of the frog.

> **Exercise 6.2 [Practice] Waterbomb Cube**
>
> Assuming the starting square of the waterbomb crease pattern in Figure 6.4(b) and Figure 6.5(a) is 1×1, what is the edge length of the cube in Figure 6.5(b)?

---

[1] This condition does not play a role for us until Sections 6.4 and 7.4.
[2] However, typically creases are created during the folding process which are later unfolded flat. Several flattened creases are evident in Figure 6.4(d).

6.3.1 Molecules

Accepting that uniaxial bases that project to a given stick figure are quite useful in origami design, the question on which we'll focus is: How to generate the crease pattern for the uniaxial base from the stick figure? In Figure 6.4, how to move from (a) to (b)?

It turns out that every uniaxial base can be constructed from a handful of ***molecules*** via the circle/river method. These molecules are something like origami gadgets, permitting the construction of complex uniaxial bases by decomposing them into molecules, and gluing the molecules together. We will describe four molecules: the rabbit ear molecule, the waterbomb molecule, the four-circle quadrilateral molecule, and the sawhorse molecule. One difference between molecules and uniaxial bases is that the boundary of a molecule is not (usually) the whole square of paper, but instead a convex[3] polygon inside the square.

6.3.2 Rabbit Ear Molecule

We start with the simplest molecule, a triangle that folds to a base shape whose shadow is a stick figure "trident" T: three edges incident to one central degree-3 node. The folded shape is shown in Figure 6.6, with the associated animation (linked in the caption) illustrating the folding from the flat triangle to the 3D uniaxial base. The triangle crease pattern is called a "rabbit ear" molecule, apparently because it was a component of a now-lost rabbit folding.

It is easy to check that this example satisfies the five uniaxial-base conditions: with flattened flaps it is uniaxial; the triangle perimeter lies along the axis; each flap hinges on a vertical crease; and the shadow is the stick figure.

Starting from the given stick figure tree T, how can the creases and M/V assignments be found to project to T? The answer is: By finding the ***incenter*** x of the triangle abc; see Figure 6.7(a). That point x is the center of the largest circle inside the triangle. This circle is tangent to all three sides, with perpendicular segments from x to the tangent points. Those three perpendiculars become the V-fold hinge creases, while the angle bisectors ax, bx, cx become the M-fold ***ridge creases***. See Exercise 6.3 for why the bisectors meet in a point.

> **Exercise 6.3 [Understanding] Bisectors Meet at Incenter**
>
> Prove that the three angle bisectors of a triangle meet at its incenter.

For the design to be uniaxial, the triangle must fold flat. The M/V crease assignment in Figure 6.7(b) is one of three ways to allow it to fold flat. See Exercise 6.4.

---

[3] For convexity, see ahead to Box 6.2.

6.3. *Uniaxial Bases* 93

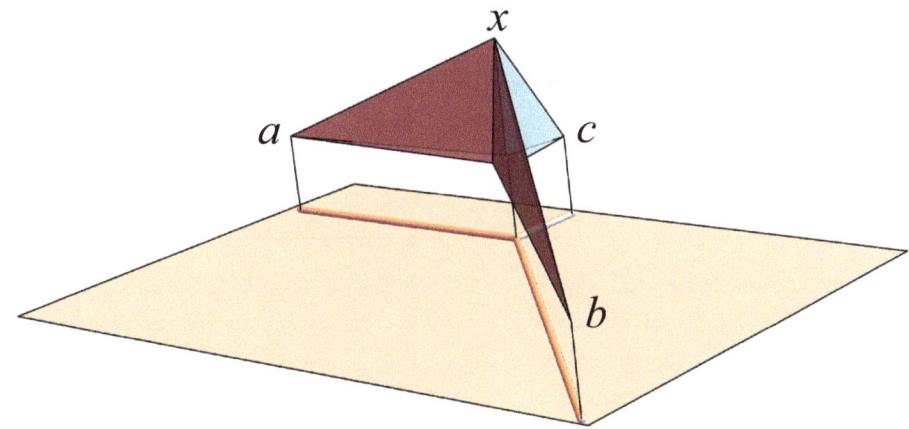

Figure 6.6 A rabbit ear molecule. The stick figure T is shown displaced downward for clarity. Animation: https://cs.smith.edu/~jorourke/MathOrigami/.

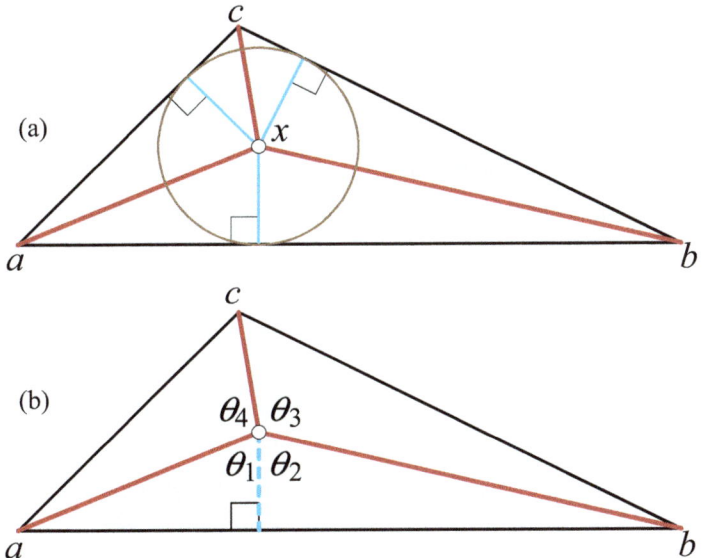

Figure 6.7 (a) x is the incenter of triangle abc. (b) The M/V creases shown allow the triangle to fold flat.

> **Exercise 6.4 [Understanding] Rabbit Ear Fold Flat**
>
> Prove that the triangle creased as in Figure 6.7(b) can fold flat, by showing that Kawasaki's Theorem 3.2 holds.

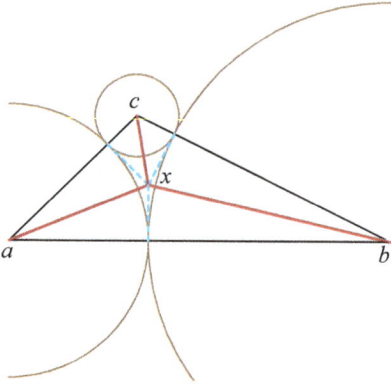

Figure 6.8 Circle packing of triangle: centers on vertices a,b,c, circles pairwise tangent.

A final remark on the rabbit ear molecule. The three perpendiculars from the incenter x are tangent to circles centered on the three triangle corners, and so the three circles are tangent to one another, verifying the fifth uniaxial-base property: see Figure 6.8. Here we start to sense why **circle packing** plays such an important role in origami.

6.3.3 Waterbomb Molecule

We've already seen in Figure 6.4(c) that the waterbomb base has four flaps projecting to a degree-4 stick figure tree with four equal-length edges, as in Figure 6.9. This is a molecule, but not a flexible one because of the equal edge length condition. Nevertheless, it is an important special case, and will even play a role in self-folding origami in Chapter 9. Again, it is easy to check the five uniaxial-base conditions.

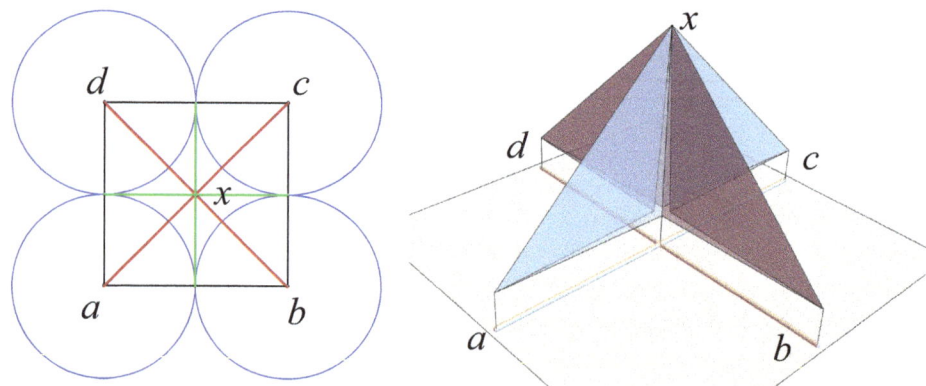

Figure 6.9 Waterbomb base projecting into degree-4 stick figure tree.

6.3. Uniaxial Bases

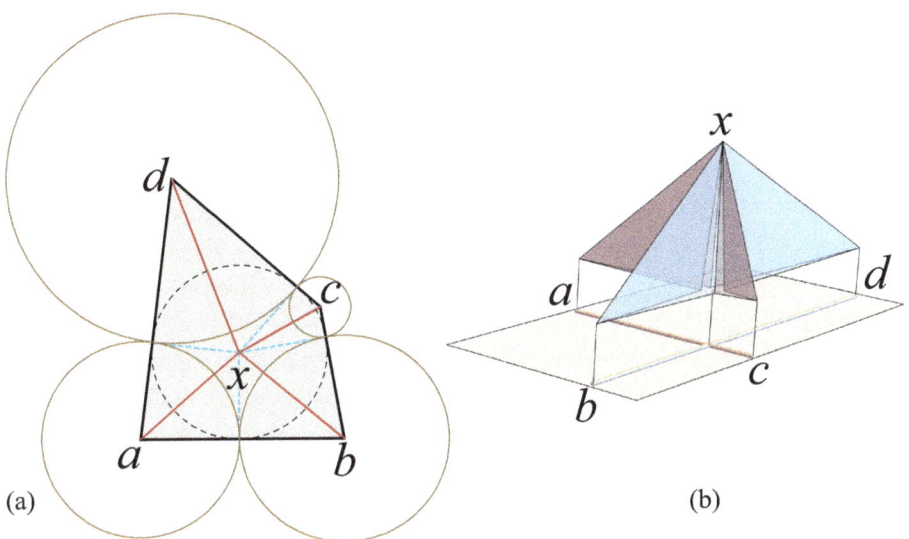

Figure 6.10 Four-circle quadrilateral molecule.

6.3.4 Four-Circle Quadrilateral Molecule

A more common situation is a need for a degree-4 stick figure tree but with different, specific edge lengths. How could a base be constructed to project to such a stick figure?

We start with a possibly irregular quadrilateral piece of paper to be situated within a complete uniaxial base later in the design process. We call this a *four-circle quadrilateral molecule*. It can be achieved by a quadrilateral whose corners are centers of mutually tangent circles, as shown in Figure 6.10(a). These four-circle quadrilaterals are special in that the four angle bisectors (red) meet at a point x, a property enjoyed by every triangle but only by these special quadrilaterals. The perpendicular from x to ab becomes a valley folded vertical hinge of the a-flap, and similarly for all four perpendiculars.

These special quadrilaterals are also known as *tangential quadrilaterals* because all four edges are tangent to an inscribed circle C. This is analogous to the inscribed circle in Figure 6.7.

Let x be the center of the circle C. The segments from the four tangent contact points on C to x are perpendicular to the quadrilateral sides. We now show that the angle bisectors meet at x as well.

Consider the angle bisector at a. As illustrated in Figure 6.11, the classical Two Tangents Theorem shows that segment ax bisects the angle at a (and at x).

> **Exercise 6.5 [Practice] Two Tangents Theorem**
>
> Prove that triangles axb' and axd' in Figure 6.11 are congruent.

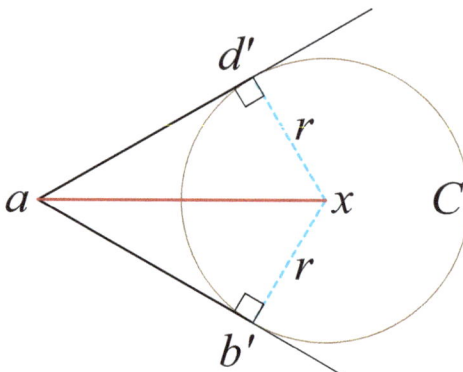

Figure 6.11 Two Tangents Theorem: Triangles axb' and axd' are congruent.

So knowing that the quadrilateral is tangential allows the four circles centered at the vertices to each be tangent to a pair of radial perpendiculars, as illustrated in Figure 6.10(a). Note that the stick figure edge lengths in Figure 6.10(b) are the radii of the four circles. The (red) bisectors ax, bx, cx, dx become M-fold ridge creases in the molecule folding, while the four (blue) radial perpendiculars from x become the four V-fold hinge creases, as shown in Figure 6.10(b).

Pitot's Theorem We know that four-circle quadrilaterals are tangential. A French engineer Henri Pitot discovered in 1725 a necessary condition for a quadrilateral to be tangential.

> **Theorem 6.1 Pitot's Theorem**
>
> In a tangential quadrilateral, the sum of the lengths of opposite sides are equal. The converse holds as well: If a convex quadrilateral has pairs of opposite side lengths equal, then it is tangential.

We'll pause our inventory of molecules and prove this attractive theorem.

The proof of Pitot's Theorem in the forward direction (tangential implies equal sums) is almost immediate from Figure 6.12. Here, A, B, C, D are the radii of the circles centered on the quadrilateral's vertices a, b, c, d respectively. Then the bottom and top side lengths sum to $A + B + C + D$, while the right and left side lengths sum to the equal quantity $B + C + D + A$. Returning to Figure 6.10(b), the sum $A + B + C + D$ is the sum of the lengths of the edges of the stick figure.

> **Exercise 6.6 [Understanding] Quadrilateral → Triangle**
>
> Provide an argument to show that every quadrilateral except a rectangle has a pair of opposite sides that can be extended to form a triangle.

6.3. Uniaxial Bases

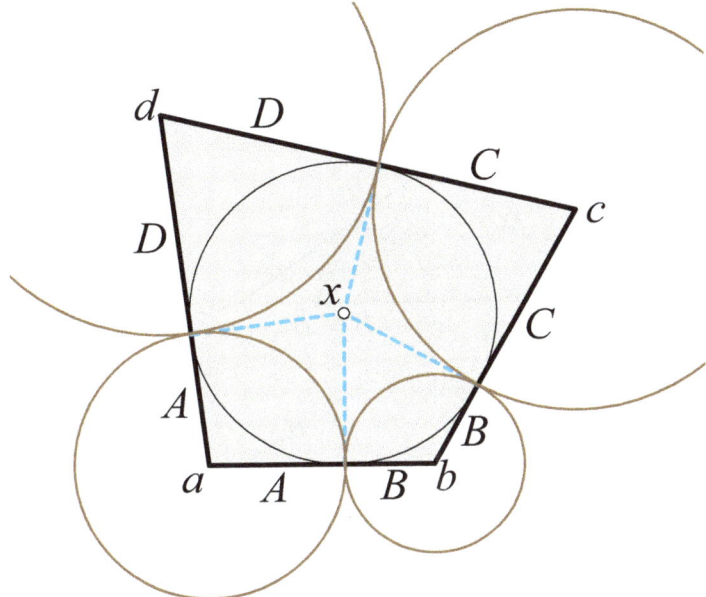

Figure 6.12 Proof of Pitot's Theorem: Equal sums of opposite side lengths.

The converse of Pitot's Theorem is more difficult to prove. It was proved by the Swiss mathematician Jakob Steiner a century after Pitot's proof, in 1846.

We prove the theorem in the contrapositive form: If the quadrilateral is not tangential, then the pairs of opposite sides do not have the same length. See Box 6.1.

> **Box 6.1 Contrapositive**
>
> Pitot's Theorem has the (short-hand) form: Tangential ⇒ Sums-equal. So the converse is Sums-equal ⇒ Tangential. The contrapositive of this negates the terms and reverses the implication: (not Tangential) ⇒ (not Sums-equal).
>
> To see this is logically equivalent, assume you want to prove that $a \Rightarrow b$ (in our case, a = Sums-equal ⇒ b = Tangential), and you've proven that $\neg b \Rightarrow \neg a$, where $\neg$ means "not." If a holds, then it cannot be that $\neg b$ holds, because $\neg b \Rightarrow \neg a$, contradicting that a holds. So if a holds, b must hold. Which proves that $a \Rightarrow b$.

We first handle the case where the quadrilateral is a rectangle. Then the side length sums imply the rectangle is a square which is tangential to the largest inscribed circle.

Now let $Q = abcd$ be a quadrilateral that is not a rectangle. Because Q is not a rectangle, we can apply the result of Exercise 6.6 to extend the opposite edges ab and dc to form a triangle $T = atd$ with its incircle centered on x. Our

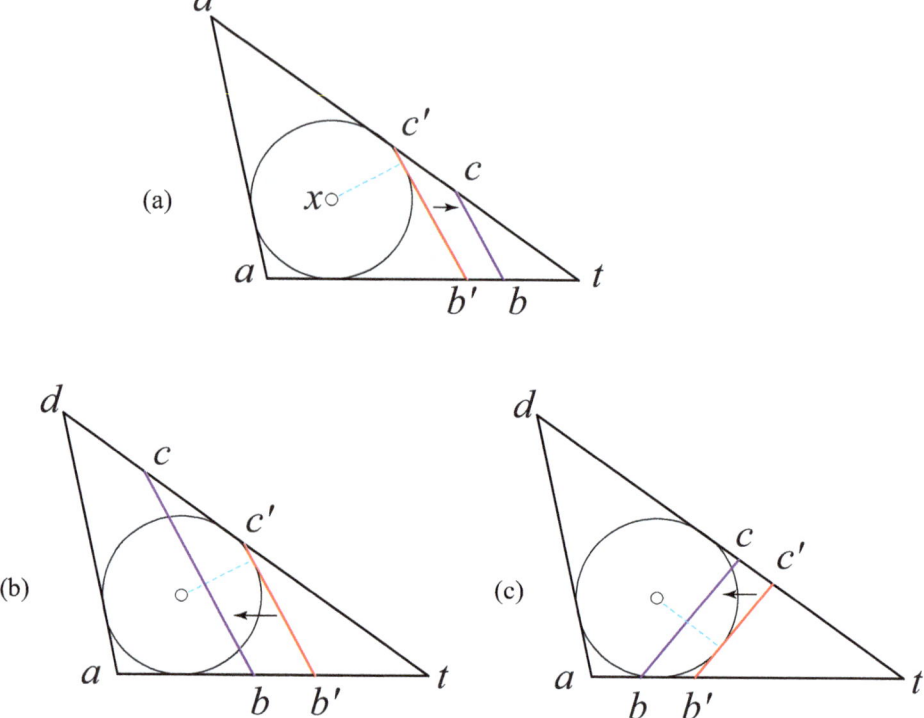

Figure 6.13 Proof of converse of Pitot's Theorem. Q is shaded quadrilateral. Q' is tangential. (a) Segment bc falls outside incircle. (b,c) Segment bc cuts into the incircle.

premise is that Q is not tangential, so it cannot be that bc is tangent to the incircle.

Let $b'c'$ be the chord tangent to the incircle and parallel to bc, marked red in Figure 6.13. The quadrilateral $Q' = ab'c'd$ is tangential, so the sums of lengths of opposite sides are equal: $A+B'+C'+D$, using the notation from Figure 6.12.

There are only two possibilities: bc is outside Q', or it cuts into Q'. Let ab be horizontal as in Figure 6.13, and call the length of ab the *bottom* length, cd the *top* length, and bc and da the *right* and *left* lengths respectively. In the first case, in Figure 6.13(a), sliding $b'c'$ rightward and parallel to itself to match bc increases *bottom*+*top*, and shortens *right* while *left* remains the same. Therefore the opposite lengths of Q separate and cannot be equal.

In the second case, Figure 6.13(b,c), sliding $b'c'$ leftward and parallel to itself to match bc decreases *bottom*+*top*, and lengthens *right* while again *left* remains the same. So again the opposite lengths separate and cannot be equal.

We have established that if the quadrilateral is not tangential, the sums cannot be equal. So if the sums are equal, the quadrilateral is tangential. This proves the converse of Pitot's theorem. So we have a complete characterization:

6.4. Circle/River Method

A quadrilateral is tangential *if and only if* the opposite length sums are equal. (Recall Box 3.2 concerning "if and only if.")

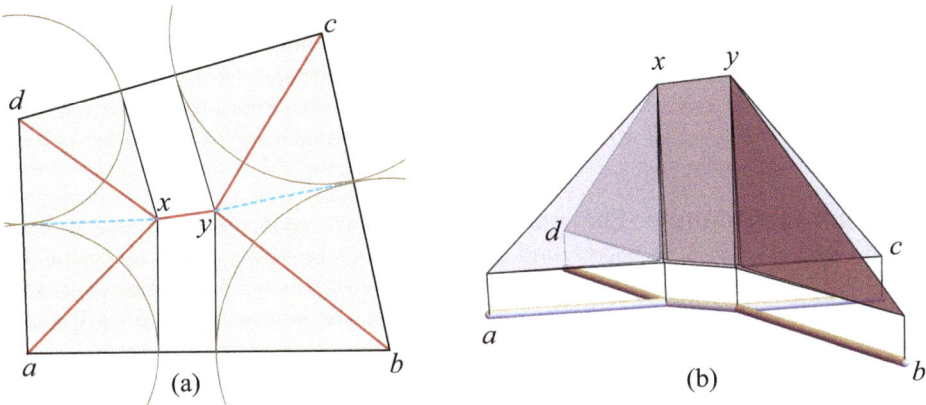

Figure 6.14 (a) Template for a sawhorse molecule. River in blue. (b) 3D rendition of folded molecule, projecting to a YY tree.

6.3.5 Sawhorse Molecule

So far we've only examined molecules with either degree-3 or degree-4 stick figure star graphs. The next simplest stick figure is the "YY" graph, two degree-3 nodes connected by a single ***internal edge***, an edge that is not incident to a leaf. Figure 6.14(a) shows a nontangential quadrilateral that can be folded to a uniaxial base by inserting a ***river*** to match the internal stick figure edge xy. This is known as the ***sawhorse*** molecule. The four circles at the corners of the quadrilateral are separated by the river, which has constant width $|xy|$, but turns to be perpendicular to both quadrilateral edges ab and cd. Inserting a river for internal stick figure edges is a general need, as we'll see in the next section.

> **Exercise 6.7 [Practice] Nontangential**
>
> Argue that the specific quadrilateral in Figure 6.14(a) is not tangential, i.e., not a four-circle quadrilateral.

6.4 Circle/River Method

Often there is a need for further molecules. For example, the stick figure for a four-circle molecule has length constraints on its four legs, as we know from Pitot's Theorem 6.1 and saw in Figure 6.10. If the four stick figure leg lengths don't satisfy those constraints, other molecules are needed. Fortunately, just five molecules suffice to build a base projecting to any stick figure: three we've discussed—rabbit ear, four-circle quadrilateral (and its specialization, the waterbomb molecule), the sawhorse—and two we opt not to detail—the arrowhead and gusset quadrilateral molecules. In this sense, the circle/river method is **universal**: A uniaxial base can be constructed for any given stick figure, no matter how complicated. We will not prove this claim, but instead describe the circle/river method at a high level, tracking two examples, one simple, one complex.

Pentagonal Example The stick figure for this example is given in Figure 6.15(a). It has five leaves and two internal edges, neither connecting to a leaf. The first step in the circle/river method is to "pack" circles onto the paper. Each leaf is assigned a circle centered on the leaf node, with radius the

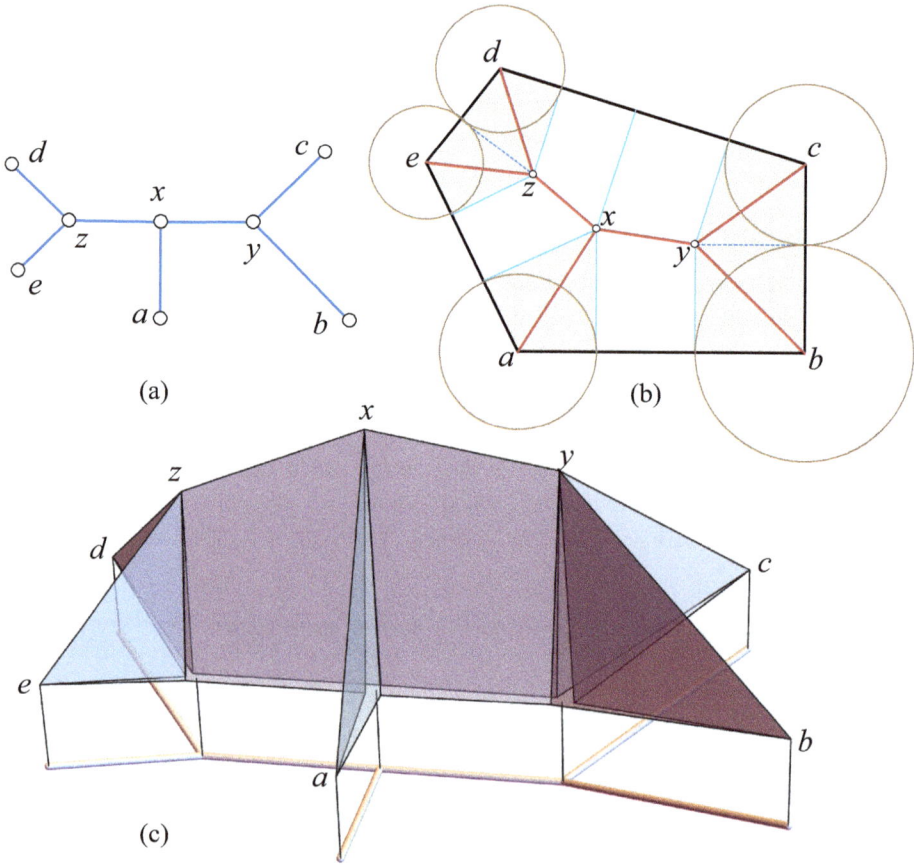

Figure 6.15 (a) The given stick figure. (b) Crease pattern that folds to a uniaxial base. (c) The base folded, projecting to the stick figure.

6.4. Circle/River Method 101

length of the stick figure edge incident to that leaf. The two internal edges are assigned rivers of appropriate widths, in this case identical widths (analogous the single river covering a single internal edge in Figure 6.14). The next step is to partition the interior of the figure into molecules, and from these derive a crease pattern. Here there is artistic freedom, and we opt to skip this step, and just claim that the pattern in Figure 6.15(b) suffices to fold to the uniaxial base in Figure 6.15(c). (Extra paper from the initial square that lies outside the pentagon can easily be folded under.) Note that again the two rivers are perpendicular to the entrance/exit edges of the pentagon.

Ten-Leaf Stick Figure Lastly we examine the impressively complex example of a 10-leaf stick figure in Figure 6.16. This example shows why the rivers are so named: They meander around the circles. Again every leaf node becomes the center of a circle, with radii corresponding to the stick figure edge lengths. The

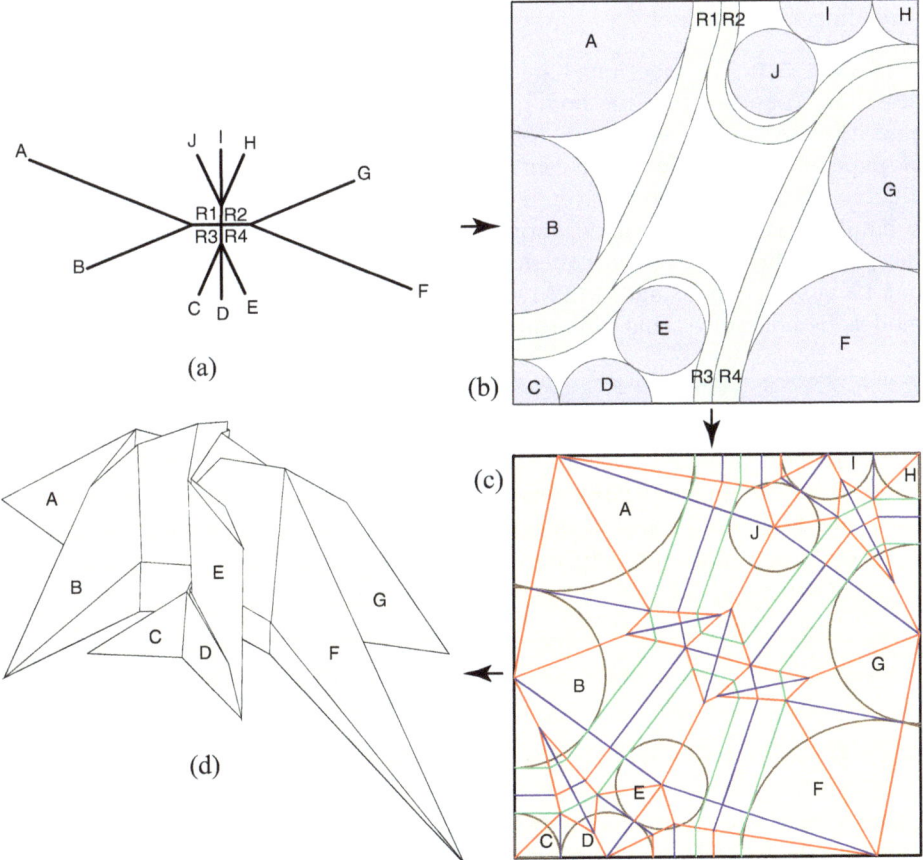

Figure 6.16 (a) Stick figure. (b) Circle/river diagram. (c) Crease pattern. (d) Folded uniaxial base. [Reprinted by permission of Arijan Abrashi, https://abrashiorigami.com/.]

four rivers (R1, R2, R3, R4) correspond to stick figure internal edges just as did the one river in Figure 6.14 and the two rivers in Figure 6.15(a). And again the rivers are perpendicular to their entry/exit edges.

Now the partition of the interior into molecules is more complicated, with some freedom for the designer. Designer choices in this case lead to the intricate crease pattern in Figure 6.16(c). Notice that centers of the circles D,E,F form a rabbit ear molecule, although the E and F circles are tangent to rivers R3 and R4 rather than to one another. Circles A,J,I form a symmetric molecule. Several other molecules can be identified in the crease pattern. It is here that the fifth uniaxial-base property plays a crucial role, ensuring that adjacent molecules are compatible, aligning tangents, a point which we'll revisit in Section 7.4. Finally, the 10 flaps in Figure 6.16(d) can all be collapsed to be uniaxial.

Remember this uniaxial base is just the first design step. The origamist now folds the flaps and other parts of the base (likely folding down overly tall flaps) to achieve some desired model—in this case, perhaps a four-legged, two-handed, three-fingered creature!

TREEMAKER We mentioned in Section 6.1 that we are stopping short of describing Lang's "tree method," which is implemented in his TREEMAKER algorithm. Although universal, the circle/river method can be quite wasteful of paper—witness the large central gap in Figure 6.16(b). Among several differences with the circle/river method, one feature of TREEMAKER is that it employs a technique called "nonlinear optimization" to pack the design in the paper efficiently. A consequence is that the algorithm is exponential-time, but Lang's success using TREEMAKER to design remarkably intricate origami models demonstrates that this running time is not a practical impediment.

> **Box 6.2 Convexity**
>
> A *convex* shape S is one for which the segment that connects each pair of points of S lies wholly inside S, i.e., is nowhere exterior to S. So S has no "dent," because a segment connecting two points straddling a dent goes exterior. ***Nonconvex*** shapes—e.g., nonconvex polygons (Section 7.3.2) or nonconvex polyhedra (Chapter 10)—often lead to greater complexity and more difficult proofs.

6.5 Straight Skeleton

There is a deep connection between uniaxial bases and the molecules that compose them, and an important structure from discrete geometry known as the ***straight skeleton***. Start with a convex polygon (Box 6.2), and steadily shrink it by moving its edges inward parallel to themselves, all at the same rate. Figure 6.17 shows this process for four of the shapes we've studied in

6.5. Straight Skeleton

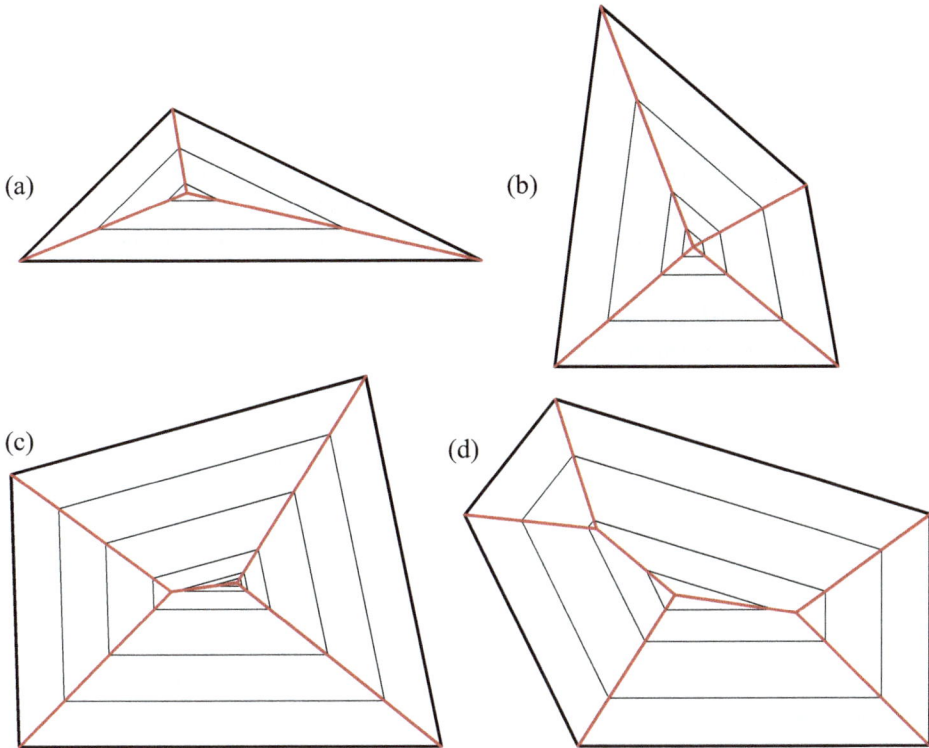

Figure 6.17 Straight skeletons (red). (a) Figure 6.6. (b) Figure 6.10. (c) Figure 6.14. (d) Figure 6.15.

this chapter: three molecules and a pentagon. The traces of the polygon's vertices during the shrinking process are precisely the ridge creases of the folded molecules!

For convex polygons, the straight skeleton is the same as the ***medial axis*** of the polygon, a central skeletal structure studied originally for its relevance to understanding biological shapes. For nonconvex polygons, the two concepts differ. We will see the role the straight skeleton plays in the Fold & 1-Cut Theorem 7.1 in the next chapter, and how this connects back to uniaxial bases.

> **Exercise 6.8 [Challenge] Straight Skeleton**
>
> (a) How many edges are in the straight skeleton of a convex hexagon?
>
> (b) Generalize to a convex n-gon.

6.6 Technical Notes

Sec. 6.1: Introduction Throughout this chapter, I rely on Robert Lang's 758-page masterwork, *Origami Design Secrets* (Lang 2012). Arijan Abrashi's website is also a great resource: https://abrashiorigami.com/.

When Robert Lang displayed a folded deer (Figure 6.1) at his invited *Symposium on Computational Geometry* talk (Lang 1996), the community was astonished and became immediately hooked on the mathematics of origami.

Sec. 6.5: Straight Skeleton The medial axis was introduced by Blum in 1967 and the straight skeleton by Aichholzer, Aurenhammer, Alberts, and Gärtner in 1995. See, e.g., Devadoss and O'Rourke (2025, Ch. 5).

7

Fold & 1-Cut

7.1 History: Harry Houdini

The topic of this chapter is an amazing theorem inspired by a magic trick.

More than a century ago, the great magician Harry Houdini described how to cut out a star from the center of a piece of paper by folding the paper four times and then cutting with "one stroke of the scissors." Figure 7.1 illustrates the method.

In Houdini's 1922 book, *Paper Magic*, he described it as follows:

> "To make a five-pointed star with one stroke of the scissors or a tear, ... [description of folds]. Care must be exercised to make the folds on exact lines, otherwise the points will be of unequal length and size. A little practice will enable you to make the folds and tear out a perfect star in ten to twenty seconds."

It is clear from this description that he started with a blank piece of paper! It might have required "a little practice" for Houdini, but even with the star predrawn on the paper it takes me much more than a minute. To fold "and tear out a perfect star" from a blank sheet that quickly would indeed appear to be magic.

What Houdini achieved for the star is now known as the ***Fold & 1-Cut*** process. Remarkably, the broadest generalization imaginable is now a theorem.

7.2 Theorem Statement

Theorem 7.1 Fold & 1-Cut

Given any straight-line drawing on a piece of paper, it is possible to fold the paper flat so that, with one straight scissors cut completely through the paper, all the drawn shapes are simultaneously cut out, leaving the original paper missing exactly those shapes.

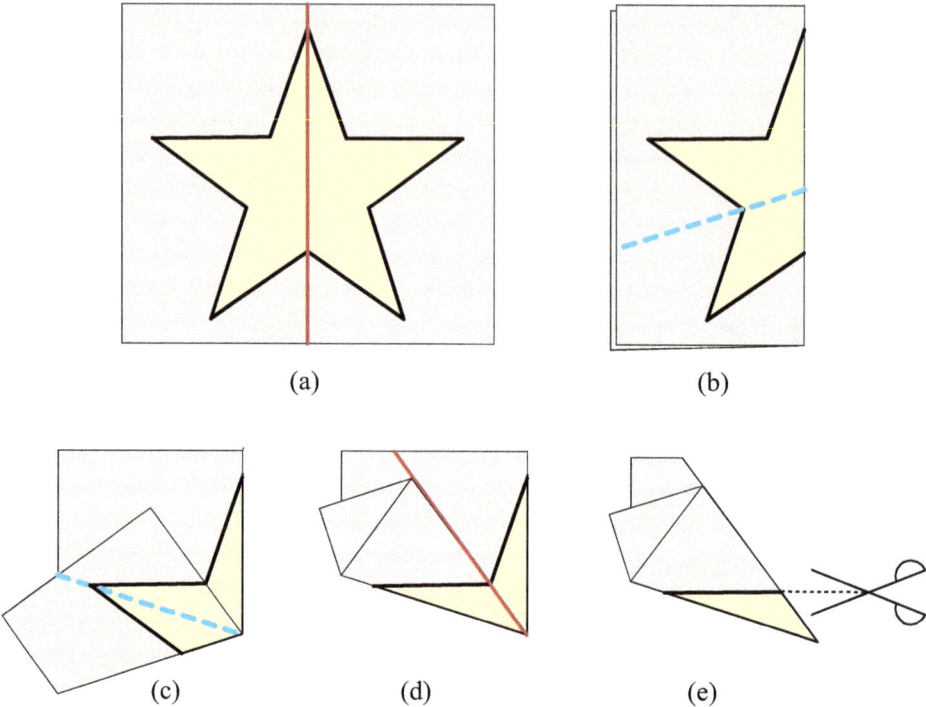

Figure 7.1 Houdini's method of folding a regular five-pointed star prepared for one-cut. As usual, M-fold: red; V-fold: dashed blue. [From O'Rourke (2011). Reprinted by permission of Cambridge University Press.]

See Figure 7.2.

> **Exercise 7.1 [Practice] 1-Cut Square**
>
> Which folds permit a single square to be cut out from the center of a larger square paper with one scissors slice?

Two Proofs There are two proofs of the theorem, both too complicated to present fully. Instead we will sketch enough of the proof steps to hopefully make the remarkable claim at least plausible if not convincing.

The first proof relies at its heart on the straight skeleton, introduced in the previous chapter, Section 6.5. This proof has the advantage of being intuitive but the disadvantage of not establishing the theorem for every possible drawing.

7.2. Theorem Statement

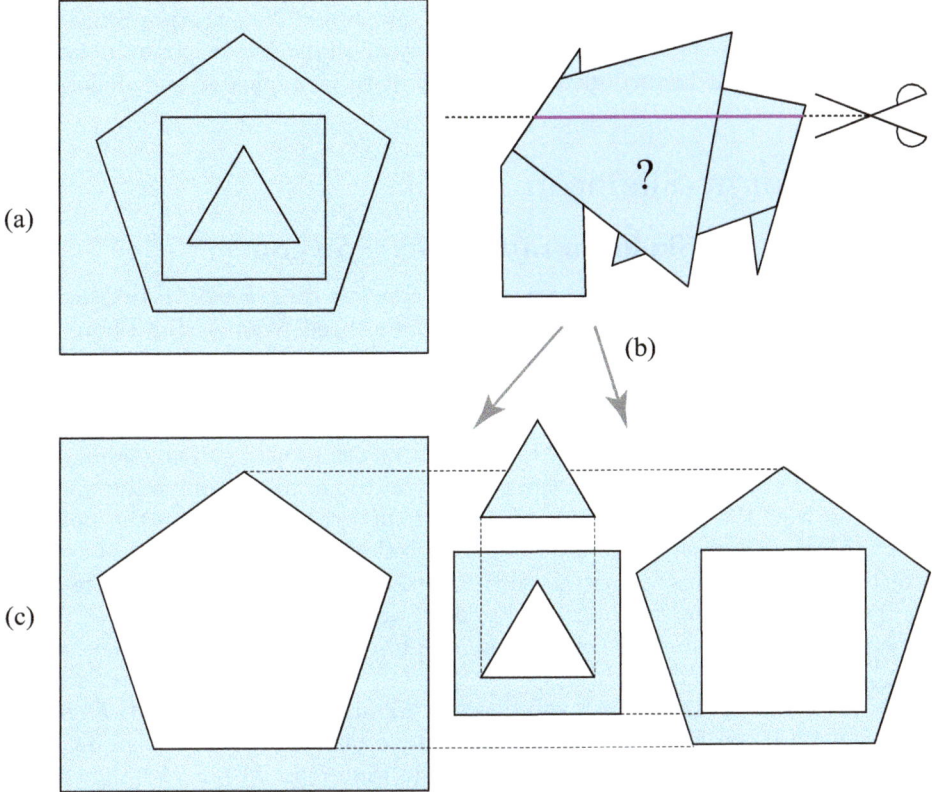

Figure 7.2 (a) A straight-line drawing. (b) Flat folding of the paper is cut straight through by scissors. (c) The shapes are cut out from the paper and fall out separately.

The second proof relies on circle packing, called in this context *disk packing*.[1] This proof has the advantage of covering every possible drawing—the full theorem statement—but the disadvantage of being quite intricate.

History of Proofs The straight-skeleton proof was discovered by Erik Demaine, Martin Demaine, and Anna Lubiw when Erik was only 17 years old. They subsequently realized there are rare drawings where the proof fails to work, as we'll describe in Section 7.3.4.

A few years later, Erik teamed with three other researchers and detailed the disk-packing proof, which circumvented the problem with those rare drawings.

---

[1] Origamists tend to prefer "circle packing" whereas mathematicians prefer "disk packing." We'll track the math literature in this chapter and use "disk packing."

But subsequently they found a flaw in this new proof. Now the proof has been repaired twice independently and is generally accepted by the community.

This thumbnail history gives some hint of the complexity and delicacy of these proofs.

7.3 Straight-Skeleton Proof

7.3.1 Straight Skeleton of Convex Polygon

Recall the straight skeleton of a convex polygon P is the trace of vertices as each edge moves inward parallel to itself at the same speed. We noted in Figure 6.17 that the skeleton edges are the ridge creases of the folded molecule. For Fold & 1-Cut purposes, the key property of skeleton edges is that they are **angle bisectors**. See Box 7.1 for definitions.

Think of the goal of folding P in preparation for 1-cutting. The folding must position the entire boundary of the polygon on top of itself along a line L, with the interior of the polygon above L and all the paper exterior to the polygon below L (or vice versa). (See Exercise 7.1 solution.) Then the 1-cut will cut out exactly P. (Note the similarity here to uniaxial bases. More on this connection in Section 7.4.) In the neighborhood of a polygon vertex v, the goal is achieved by creasing along the angle bisector at v, folding the two incident edges to lie on top of one another.

Let us revisit the four examples in Figure 6.17. For the triangle in Figure 6.17(a), by Exercise 6.3 the three angle bisectors meet at a point. The same holds for the four-circle quadrilateral in Figure 6.17(b). So for the shapes in Figure 6.17(a) and (b), the entire skeleton is composed of angle bisectors. In the sawhorse quadrilateral in Figure 6.17(c), the four angle bisectors lead to a central segment xy. But xy is itself a subsegment of a bisector of the top and bottom edges. For the pentagon in Figure 6.17(d), there are five angle bisectors and two central segments, xy and xz in Figure 6.15(b). One can see that xy is a subsegment of the bottom and top edges ab and cd, while xz is a subsegment of the top and left edges cd and ae. So each straight-skeleton edge in these examples is an angle bisector, and therefore ideally suited for aligning the two edges that it bisects.

Box 7.1 Angle Bisectors

Let e_1 and e_2 be the two polygon edges incident to a vertex v, forming an angle α there. Then the **angle bisector** is a ray from v toward the interior of the polygon that makes angles α_1 and α_2 with e_1 and e_2 respectively, with $\alpha_1 = \alpha_2 = \alpha/2$.

If e_1 and e_2 are two nonconsecutive edges of a polygon, their angle bisector is a line L. If e_1 and e_2 are parallel, L is midway between them. If they are not parallel, then the extensions of e_1 and e_2 toward one side meet at a point x. Then L is the line bisecting the angle at x.

7.3. Straight-Skeleton Proof

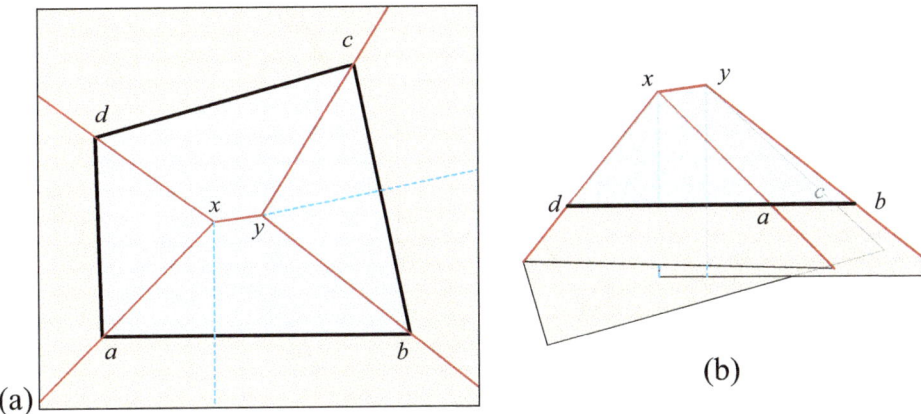

Figure 7.3 (a) Sawhorse quadrilateral with straight skeleton edges and two perpendiculars. (b) Flat folding prepared for 1-cut.

Perhaps it is not surprising now that the first step to fold and 1-cut a convex polygon is to crease every straight skeleton edge, both interior to the polygon and extended exterior to the boundary of the paper. This aligns the angle bisectors, but these creases are not enough to fold flat. Consider the sawhorse polygon from Figure 6.17(c), redrawn in Figure 7.3(a). The two internal skeleton vertices x and y each have three incident M-folds. Another crease is needed to satisfy Maekawa's Theorem. Each of these creases can be provided by two valley **perpendiculars**, one incident to x and perpendicular to edge ab, and another incident to y and perpendicular to edge bc. Folding along a perpendicular will align the two portions of the boundary on either side of the perpendicular, and so will not destroy what is achieved by the angle bisectors.

One can see there are three choices for each perpendicular, one into each of the three skeleton faces incident to the skeleton vertex. Indeed Figure 6.14 in the previous chapter used a perpendicular from x to edge da rather than from x to edge ab as here. All choices result in different flat foldings, but always with the boundary aligned and ready for a 1-cut, as in Figure 7.3(b).

Although we haven't formally proved this intermediate lemma, even with its limitations it is already impressive.

Lemma 7.1 Fold & 1-Cut for Convex Polygon

A convex polygon can fold flat and be 1-cut out by a crease pattern consisting of M-folds for the straight skeleton edges and V-fold perpendiculars for each skeleton vertex, with all creases extending exterior to the polygon out to the paper boundary.

Straight skeletons generically have degree-3 vertices (as in Figure 6.17), but higher degrees are possible. For example, the straight skeleton of the

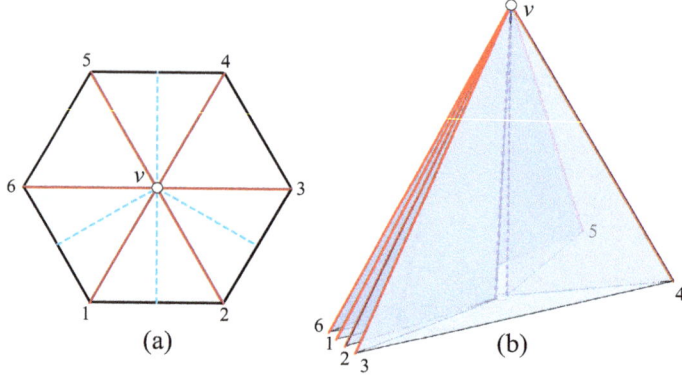

Figure 7.4 (a) Four perpendiculars added to degree-6 vertex v. Surrounding paper not shown. (b) Nearly flat folding of hexagon, when all six hexagon edges are superimposed.

regular hexagon in Figure 7.4(a) has a degree-6 vertex at its center v. To satisfy Maekawa's Theorem 3.1, four perpendiculars are needed to fold flat; see Figure 7.4(b).

Much remains to reach the full Theorem 7.1, starting with nonconvex polygons.

7.3.2 Straight Skeleton for Nonconvex Polygons

Figure 7.5 shows the straight skeleton for a nonconvex pentagon. In the shrinking process, when edge ab collides at point z with the vertex d trace, the skeleton splits at z, in this case into two triangles centered on x and y. The shrinking then proceeds on the two triangles independently.

Similar to Lemma 7.1, we add perpendiculars to each skeleton vertex. Then we crease each skeleton edge and each perpendicular, assigning M/V folds so that the pattern folds flat, and we are prepared for the 1-cut: see Figure 7.6.

Some subtleties concerning perpendiculars remain, however, which we will see in Section 7.3.4. Before that we first study the straight skeleton in more detail.

> **Exercise 7.2 [Practice] Folded Layers**
>
> When the pentagon in Figure 7.6 is folded flat for 1-cut, what is the most layers of paper the scissors must cut through?

A tangential but attractive aspect of the straight skeleton connects to architecture. If one imagines the edges of the polygon not only moving inward at the same rate, but lifting vertically at the same rate, the result is a ***straight-skeleton polyhedron*** as illustrated in Figure 7.7. The faces of this polyhedron

7.3. Straight-Skeleton Proof

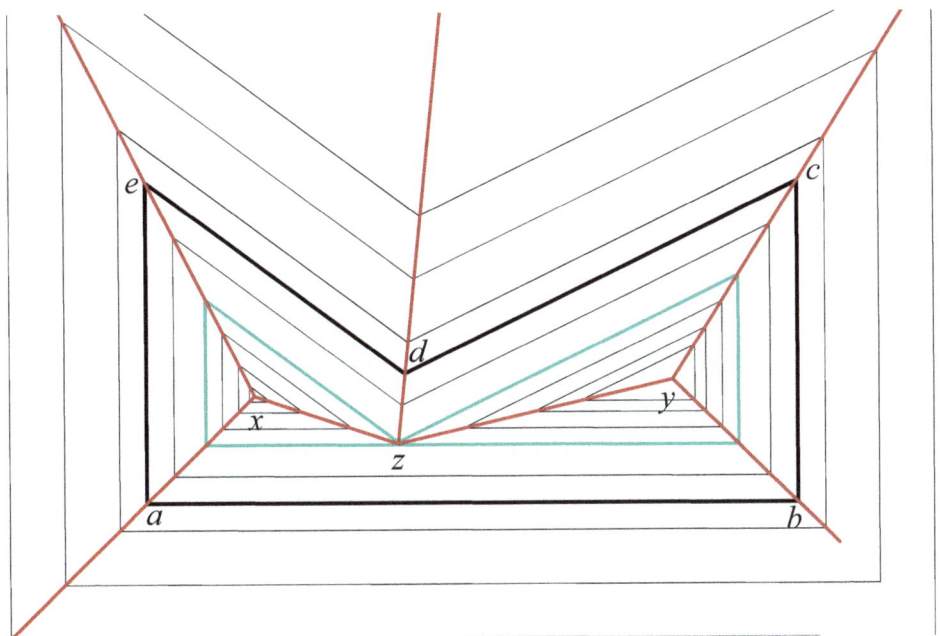

Figure 7.5 Straight skeleton of a nonconvex pentagon. Split event at z.

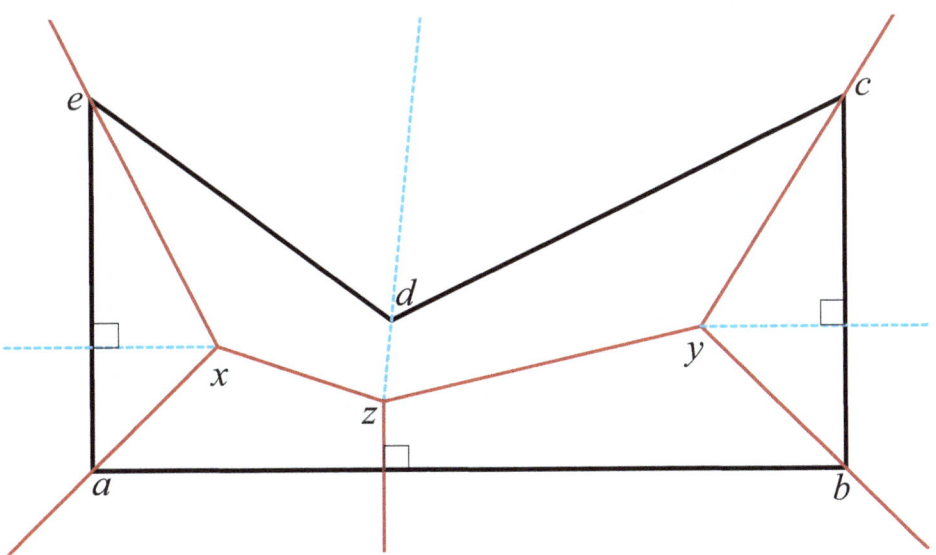

Figure 7.6 M/V creases to fold flat and 1-cut the nonconvex pentagon.

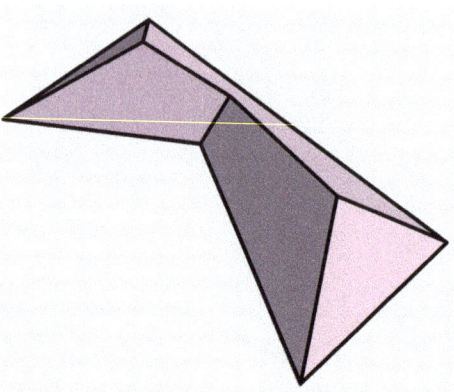

Figure 7.7 Polyhedron corresponding to Figure 7.5.

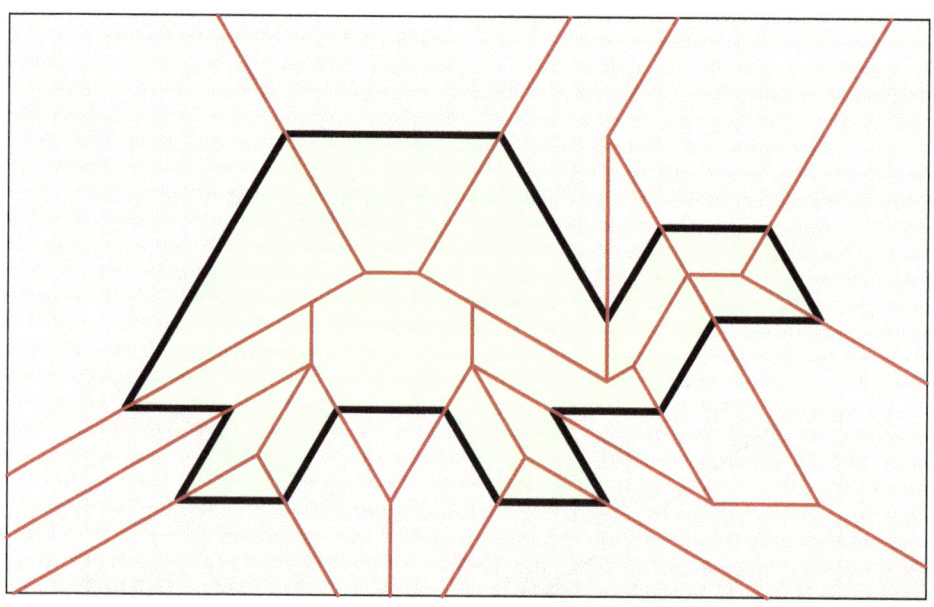

Figure 7.8 Straight skeleton of turtle polygon. [From Demaine and O'Rourke (2007). Reprinted by permission of Cambridge University Press.]

all have slope 1, and form an attractive roof design, already discussed in the nineteenth century.

7.3.3 Structure of Straight Skeltons

The straight skeleton for a more complicated nonconvex polygon is shown in Figure 7.8. Recall we observed that the interior skeleton segment xy in

7.3. Straight-Skeleton Proof

Figure 7.3(a) was a subsegment of an edges bisector. This crucial property for Fold & 1-Cut remains true for any straight skeleton. We will not prove this claim, but use the turtle skeleton to illustrate it.

A *face* of a straight skeleton is a region surrounded by skeleton edges or edges of the paper boundary. If a polygon has n edges, its straight skeleton has n faces, $n = 17$ in Figure 7.8.

> **Lemma 7.2 Edge–Face Correspondence**
>
> There is a one-to-one correspondence between edges of the polygon and faces of the straight skeleton: Each edge lies in one face of the skeleton, and each face of the skeleton contains one edge of the polygon.

Here is the key bisection lemma.

> **Lemma 7.3 Skeleton Edge Bisectors**
>
> Each edge e of the straight skeleton is shared by two faces of the skeleton, f_1 and f_2. Edge e is a subsegment of the bisector of the polygon edges in f_1 and f_2.

In Figure 7.9, e (red) is an edge of the skeleton shared by the two skeleton faces f_1 and f_2 (blue). By Lemma 7.2, each of these faces contains exactly one polygon edge, say, e_1 and e_2. The extension of those edges (dashed) meet at a point x, showing that e is a subsegment of the bisector of e_1 and e_2. Choosing any other edge of the skeleton and performing the same analysis verifies that Lemma 7.3 holds throughout.

To prepare for 1-cutting, again we will crease each skeleton edge and each perpendicular.

7.3.4 Perpendiculars

For convex polygons, the perpendiculars exit the polygon and run directlly to the paper boundary. For nonconvex polygons, a perpendicular might hit an edge of the straight skeleton, as illustrated in Figure 7.10. One can see that, to maintain the bisection property of that skeleton edge in the neck of the turtle, the perpendicular crease must reflect across the edge to match itself (lay upon itself) after folding the bisector. This reflection may be repeated until eventually the perpendicular exits the paper. Call a perpendicular that doesn't head straight to the paper boundary ***wandering***. Of course if the perpendicular were chosen in this example to rise vertically or rightward from x rather than downward, the wandering behavior would be avoided. But sometimes all choices lead to wandering.

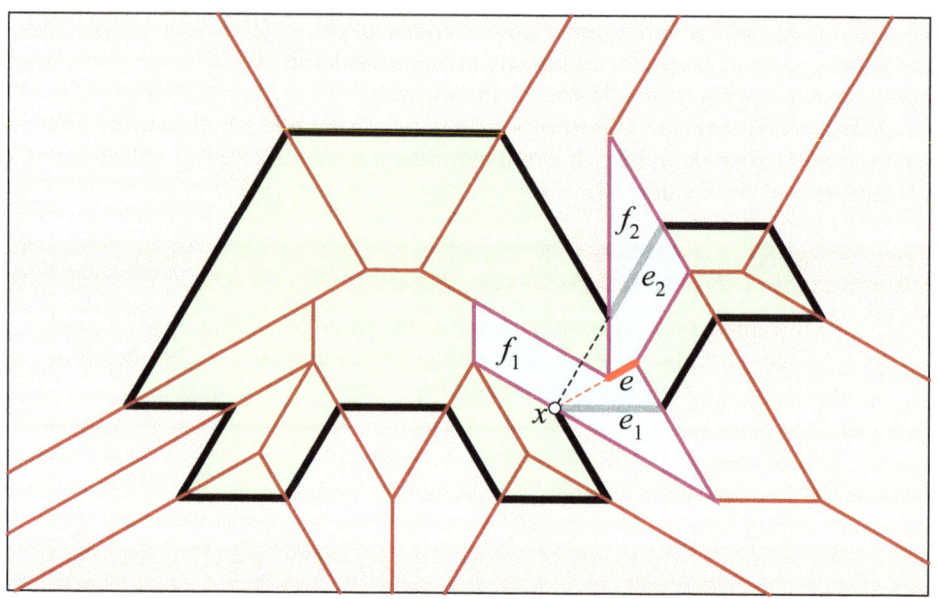

Figure 7.9 Skeleton edge e bisects polygon edges e_1 and e_2.

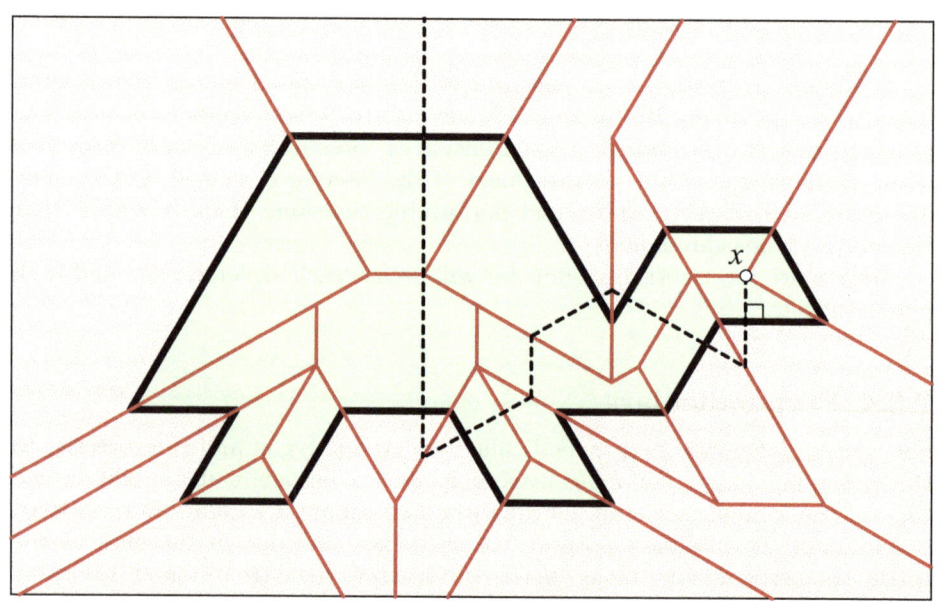

Figure 7.10 Wandering perpendicular originating at x.

7.3. Straight-Skeleton Proof

Figure 7.11 Turtle polygon boundary (black) nearly aligned for 1-cut.

> **Exercise 7.3 [Understanding] Wandering Perpendiculars**
>
> Identify a vertex of the turtle straight skeleton in Figure 7.10 whose perpendicular is necessarily wandering, i.e., does not head straight to the paper boundary.

Wandering itself is not a fatal problem, but in rare circumstances a perpendicular might never leave the paper, and instead spiral more and more densely, resulting in an unfoldable infinitely many creases. This is why the straight-skeleton proof does not handle every possible straight-line drawing. Let us call a perpendicular that avoids dense spiraling a ***finite perpendicular***.

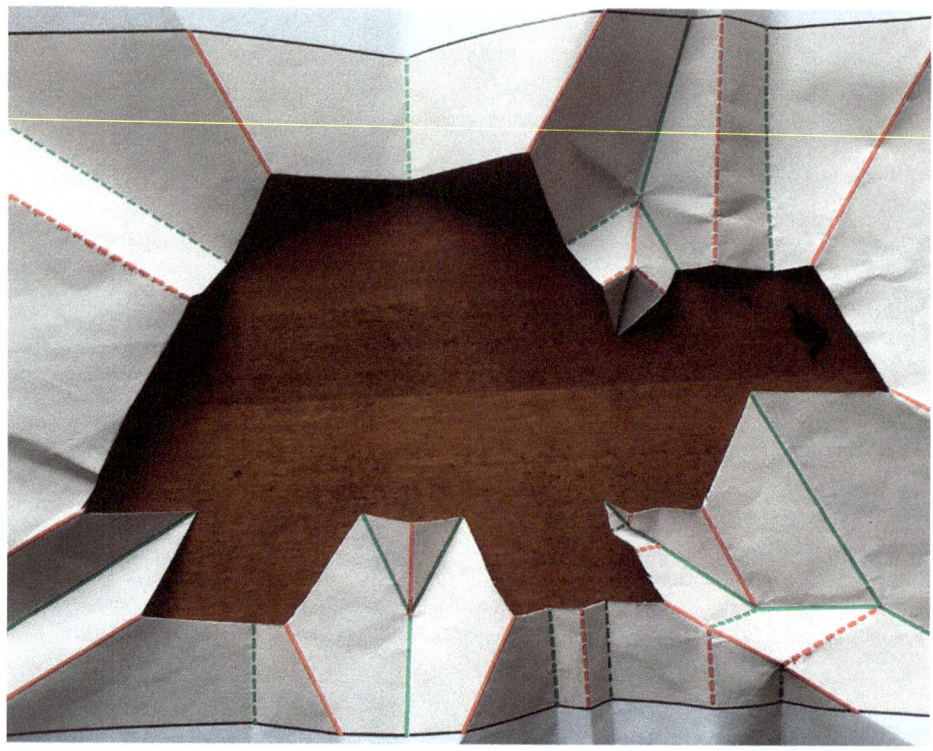

Figure 7.12 After 1-cut, turtle hole remains.

> **Theorem 7.2 Fold & 1-Cut by Straight Skeleton**
>
> If a polygon's straight skeleton can be augmented by finite perpendiculars for each skeleton vertex, then the polygon can fold flat and be 1-cut out by a crease pattern consisting the edges of the polygon's straight skeleton and those perpendiculars, with appropriate M/V assignments.

We've skipped many details, but indeed it does work: see Figures 7.11 and 7.12.

Theorem 7.2 extends to collections of polygons by defining the straight skeleton for several polygons. The end result is that Theorem 7.1 is proved for all drawings that have finite perpendiculars. However, precisely determining these drawings remains unresolved.

> **Open Problem 7.1 Finite Perpendiculars**
>
> Characterize those straight-line drawings that avoid infinite perpendiculars.

7.4. Disk-Packing Proof

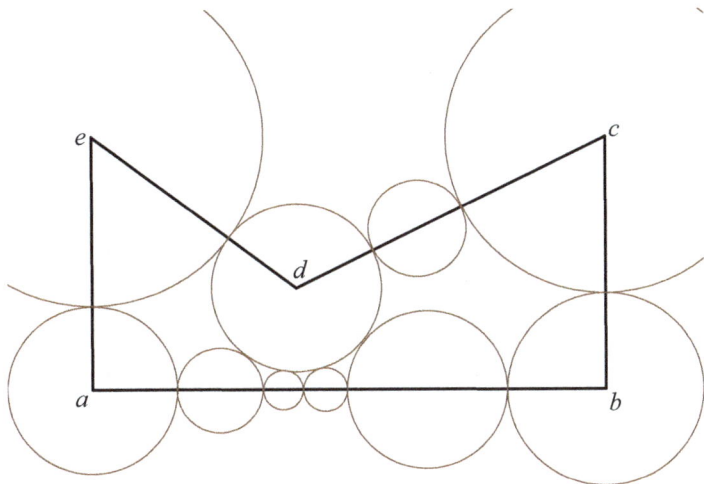

Figure 7.13 Covering vertices and edges of P with disks. Note the five-sided gap inside the pentagon.

7.4 Disk-Packing Proof

As mentioned in Section 7.2, the only full and correct proof of Theorem 7.1, covering all straight-line drawings, is the disk-packing proof. This proof is even more intricate than the straight-skeleton proof, and we'll only present such a high-level view of the proof that it borders on a caricature. Still we'll see that the proof connects nicely to unaxial molecules in Section 6.3.1.

Let P be a single polygon. We will illustrate the proof steps applied to the nonconvex pentagon in Figure 7.6. The first step is to center nonoverlapping disks on each vertex, and then cover the edges not yet fully covered with disks centered on those edges, as illustrated in Figure 7.13. The boundary of P is now covered by tangential disks.

> **Exercise 7.4 [Understanding] Covering by Disks**
>
> Given any number m, construct a quadrilateral whose edge-cover by tangential disks requires more than m disks.

Second, add more nonoverlapping disks to fill out to the edge of the paper, and—crucially—so that all gaps between disks are bounded by either three or four circle arcs. Figure 7.13 shows one five-sided gap (toward the right side), which is reduced to one three-sided gap and two four-sided gaps in Figure 7.14.

Third, connect the centers of each pair of touching disks with segments. This partitions the paper, both inside and outside P, into triangles and

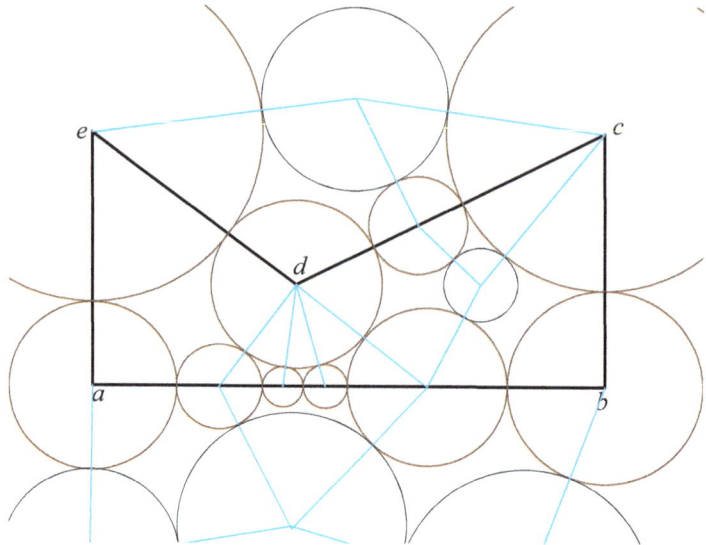

Figure 7.14 Additional disks added to cover paper. Touching disks leave only three- or four-sided holes, corresponding to triangles or quadrilaterals.

quadrilaterals. See again Figure 7.14. Because the boundary of P is comprised of touching disks, the interior of P is partitioned into triangles and quadrilaterals.

Now look at one such quadrilateral inside P: see Figure 7.15. It is a tangential quadrilateral, and so a quadrilateral molecule. We finally here see why Lang's uniaxial molecules need the condition that the four circles are tangent at "kissing points" on the quadrilateral's edges: because this ensures that an adjacent molecule's circles will be compatible. For example, the triangle to the right of db' in the figure shares with the quadrilateral two circles centered on d and b'.

Next, imagine cutting out each molecule separately, and folding it independently so that its boundary becomes uniaxial. Recall the universality claim from Section 6.4 guarantees that this is possible. Note that the molecules are either wholly interior to P or wholly exterior to P, because the boundary of P is covered by disk center-to-center connections. Arrange all the interior folded molecules to be above the common line L to each uniaxial base, and the exterior molecules to lie below L. Here several tricky details are being suppressed.

Now comes the magic: Glue all the separated molecules back together and voila!—We have aligned the boundary ready for 1-cutting.

Gluing the separated molecules back together while avoiding self-intersections is a delicate process, and the source of one of the oversights in the original proof. Matters become more complicated for multiple polygons, and even more complicated for nested polygons. Surmounting these complexities took a decade, but the final proof handles any straight-line drawing, establishing the full generality of the Fold & 1-Cut Theorem 7.1. This remains one of the deepest and most beautiful theorems in the mathematics of origami.

7.5. *Flattening Polyhedra* 119

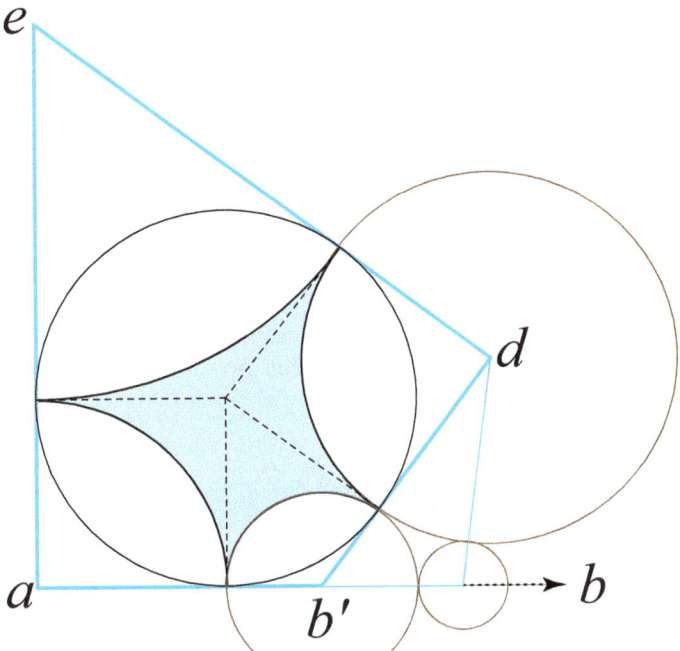

Figure 7.15 One tangential quadrilateral molecule inside the pentagon.

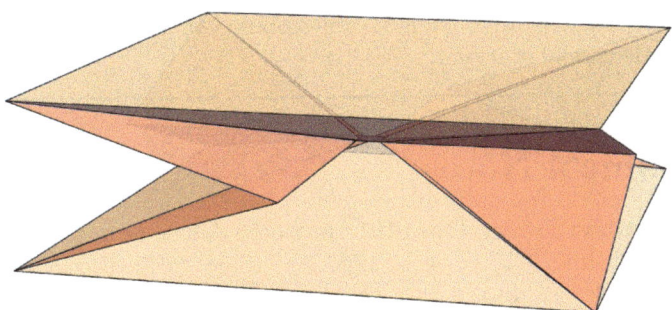

Figure 7.16 Cube flattening using "rolling" creases. Animation: https://cs.smith.edu/~jorourke/MathOrigami/.

7.5 Flattening Polyhedra

The Fold & 1-Cut process folds a 2D drawing to a 1D line. Specialize "drawing" to a polygon, and consider the natural generalization in one higher dimension: **flatten** a 3D polyhedron to a 2D plane. In general this flattening cannot be accomplished by rigid folding, but if "rolling" creases are permitted, then flattening is feasible: see Figure 7.16 and its animation. It is now known that convex polyhedra can be flattened, via two different proofs, one of which follows the straight skeleton. It remains open to extend either of these proofs to nonconvex polyhedra.

Setting aside this sense of flattening, there is a somewhat fanciful generalization of Fold & 1-Cut for a 2D polygon to Fold & 1-Cut for a 3D polyhedron. The folding of the 2D polygon's boundary to a 1D line is accomplished by folding the 2D paper along 1D creases, rotating the paper into 3D around the crease axis and returning the paper to 2D in multiple layers.

Analogously, we could imagine solid "3D paper" containing an embedded polyhedron, which we flatten to a 2D plane by rotating the solid paper in 4D about 2D "crease planes," returning the paper to 3D in multiple layers. Once the boundary of the polyhedron lies entirely in a 2D plane inside 3D, "4D scissors" slice that plane, and the flattened polyhedron drops out of the solid paper!

This generalization extends to higher dimensions (Box 7.2), but has not yet been successfully pursued.

> **Box 7.2 4D Rotations**
>
> An advanced topic: Why does a 4D rotation rotate about a crease *plane*?
>
> - A 2D rotation has a single "0D" stationary point: The center of rotation does not move.
> - A 3D rotation has a stationary line: the 1D axis of rotation.
> - A 4D rotation has a 2D stationary plane.
>
> This pattern extends to higher dimensions: An n-D rotation has an $(n-2)$-dimensional stationary subspace.

7.6 Technical Notes

Sec. 7.1: History: Harry Houdini Houdini's description in his *Paper Magic* book (Houdini 1922) is on p. 176. The most detailed expositions of the straight-skeleton proof and of the disk-packing proof are in Demaine and O'Rourke (2007, Ch. 17). The original disk-packing proof is Bern et al. (2002).

Sec. 7.4: Disk-Packing Proof The high-level proof outline here is derived largely from https://erikdemaine.org/foldcut/.

Sec. 7.5: Flattening Polyhedra The two methods of flattening any convex polyhedron via rolling creases are Itoh et al. (2012) and Abel et al. (2014). For the "fanciful generalization," see Sec. 18.2 of Demaine and O'Rourke (2007).

8

Curved-Crease Origami

8.1 Introduction

The mathematical study of curved-crease origami was initiated in the 1970s through the work of Ronald Resch and David Huffman.[1] Huffman's models in particular were influential, achieving an almost sculptural beauty despite being folded from a single piece of paper: see Figure 8.1.

Today curved-crease origami is an art form with several masterful practitioners, for example Ekaterina Lukasheva—Figure 8.2, and Erik and Martin Demaine—Figure 8.3.

The mathematics underlying curved creases falls mainly in the field of "differential geometry," well beyond the scope of this book. Moreover, this is an active area of research, with mathematical understanding advancing regularly at the frontiers. So our exploration of this topic is necessarily superficial, but nevertheless hopefully rewarding.

We start with a simple example in Section 8.2 with one of the rare curved-crease designs whose geometry is well understood. A theme we'll see in this chapter is viewing curved surfaces as approximated by traditional straight-line creases. Section 8.3 explores the best known such case: the straight-crease model of the hyperbolic paraboloid, which in fact is not a correct model. Next in Section 8.4 we introduce the notion of curvature, a concept possibly new to many readers, which then allows us to describe (without proof) three general curved-crease theorems in Section 8.5.

We close this chapter in Section 8.6 with a remarkable connection between conic curved creases and the flat-foldable degree-4 vertices studied in Chapter 5.

A note on notation: We'll use $\mathcal{C}$ for a curve in 3-space, and when needed to distinguish 2D from 3D, Γ for a curve in a plane (Section 8.4).

---

[1] Huffman is the inventor of "Huffman coding," used for example in the JPEG standard for image compression.

Figure 8.1 David Huffman *Pinwheel*, 1970s. [Used with permission of Taylor & Francis Group, from Demaine et al. (2011b); permission conveyed through Copyright Clearance Center, Inc.]

8.2 *Vesica Piscis*

One challenge to appreciating and understanding curved-crease origami is that it takes great skill to fold examples like those displayed in the previous section. So although amateurs can follow straight-crease origami instructions, the same does not hold for curved-crease origami.

An exception is folding the *vesica piscis* pattern, accessible to even unskilled folders. The term *"vesica piscis"* describes the two-circle pattern shown in Figure 8.4. (In Latin the phrase means the bladder of a fish, referring to the central lens-shape.) This pattern arises in Euclid's method of constructing an equilateral triangle on a given line segment.[2] Center circles at each endpoint of the segment, each with radii the length of the segment. Then either of the two intersections of the circles determines an equilateral triangle.

> **Exercise 8.1 [Practice] Area of Lune**
>
> For unit-radius circles, what is the area of the lune?

---

[2] Book I, Proposition 1.

8.2. *Vesica Piscis* 123

Figure 8.2 Ekaterina Lukasheva: *Mystery*, Opus T-204, 2019. [Reprinted by permission of the artist, Ekaterina Lukasheva.]

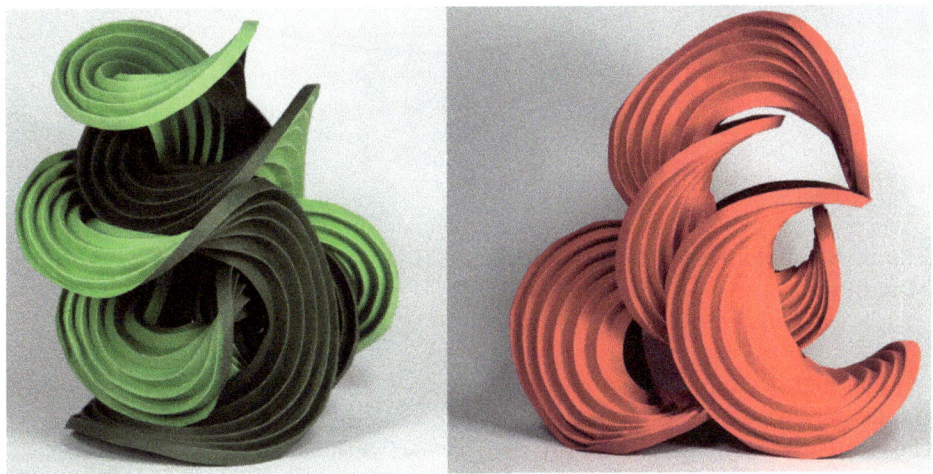

Figure 8.3 Erik and Martin Demaine: Curved-crease sculptures composed of several pieces of paper. [Reprinted by permission of the artists, Erik and Martin Demaine.]

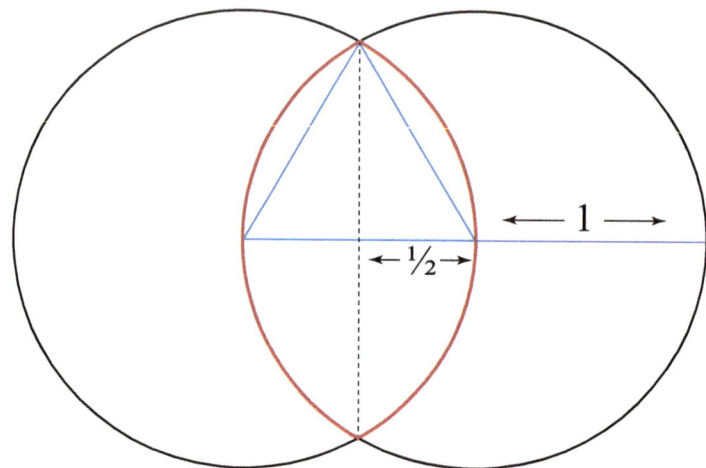

Figure 8.4 *Vesica piscis*. The red curves receive mountain-folds.

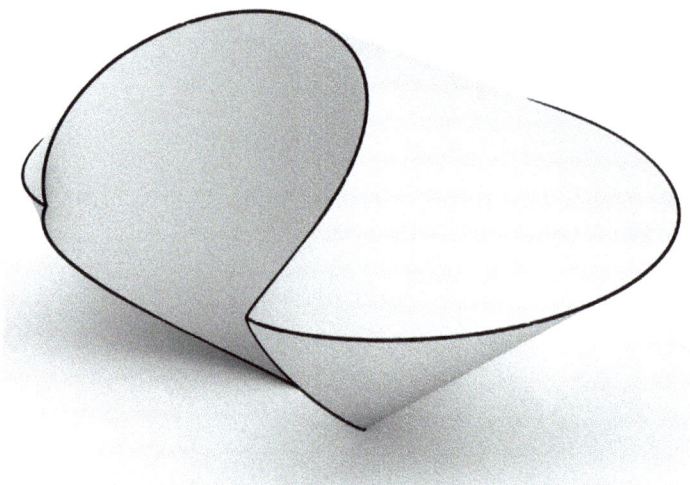

Figure 8.5 Folding of the *vesica piscis*. [Reprinted by permission of Klara Mundilova.]

The fold is simple, and I encourage readers to try it. Cut out the union of the two circles, and mountain-fold the two red circular arcs. It is best to score those arcs on the back side and then crease with your fingers. To complete the construction, glue (or tape) together the black circles. The result is the aesthetically pleasing shape shown in Figure 8.5.

As Figure 8.6 illustrates, the top/bottom halves join symmetrically along a plane. Each half is formed by a cone intersected with a cylinder. Call the curved crease in space $\mathcal{C}$. With unit-radius circles used in Figure 8.4, the cone

8.3. Pleated Hyperbolic Paraboloid

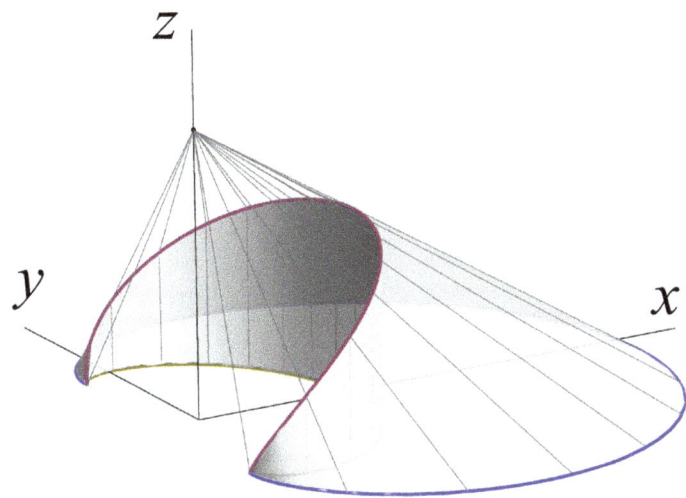

Figure 8.6 Cone with cylinder removed. [Reprinted by permission of Klara Mundilova.]

apex has height $\frac{3}{4}$ over the origin, and the highest point of $\mathcal{C}$ has height $\frac{1}{2}$. $\mathcal{C}$ lies at the intersection of the cone and the cylinder—a beautiful construction of elementary shapes!

Note the curve $\mathcal{C}$ does not lie in a plane. Despite the simplicity of the cone/cylinder description, deriving equations to describe $\mathcal{C}$ is by no means straightforward. A mathematical description of $\mathcal{C}$ is now known, but it is rather complicated, involving "elliptic integrals of the third kind"! As we'll see in Section 8.3, it is not always easy to prove the precise 3D shape that results with curved creases. The *vesica piscis* folding is a rare exception. Short of the complex proof, there is some regularity to $\mathcal{C}$ (and any curved crease) that we'll address later in Figure 8.12.

8.3 Pleated Hyperbolic Paraboloid

A parabola in the xz-plane has equation $z = x^2$, which when spun about the z-axis forms a paraboloid in 3D with equation $z = x^2 + y^2$. Changing the "+" to a "−" leads to a **hyperbolic paraboloid**, the surface with the equation $z = x^2 - y^2$.

This surface is often known as a "saddle surface";[3] see Figure 8.7(a). With the standard meshing shown, there seems little connection to origami. But the hyperbolic paraboloid can be viewed—strikingly—as constructed from two sets of crossing straight lines embedded in the surface, as shown in Figure 8.7(b). It has long been known that a pleating crease pattern consisting of concentric

---

[3] Roughly the shape of a Pringle.

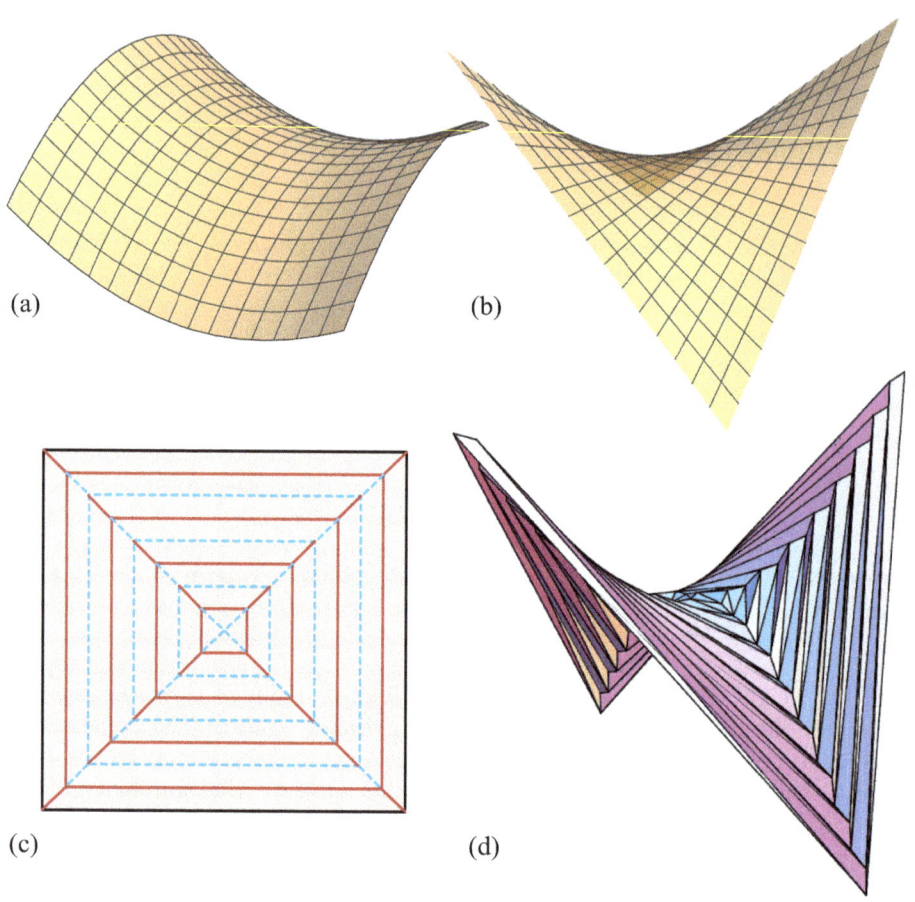

Figure 8.7 (a) Hyperbolic paraboloid. (b) Same surface but "doubly ruled." (c) Nested-squares pleat pattern. (d) Shallow diagonals split rectangles between M and V creases. [Used with permission of Springer Nature, from Demaine et al. (2011a); permission conveyed through Copyright Clearance Center, Inc.]

squares and diagonals to the center (Figure 8.7(c)) naturally forms a shape in space that approximates, and maybe equals, a hyperbolic paraboloid.

The nested-circles version in Figure 8.8(a) has been studied for at least a century, going back to the Bauhaus art school. Folding variations on this pattern lead to the spectacular paper sculptures shown earlier and in Figure 8.8(b).

It had been assumed for years that these pleated models "exist," exist in the sense that paper can be folded exactly along the crease patterns in Figures 8.7(c) and 8.8(a), with no other creases. However, a difficult 2009 proof established that the nested-squares crease pattern does not fold along those creases. There must be other creases to permit it to fold, long hard-to-discern diagonals creasing the thin rectangles, as shown in Figure 8.7(d). The dihedral angles

8.3. Pleated Hyperbolic Paraboloid

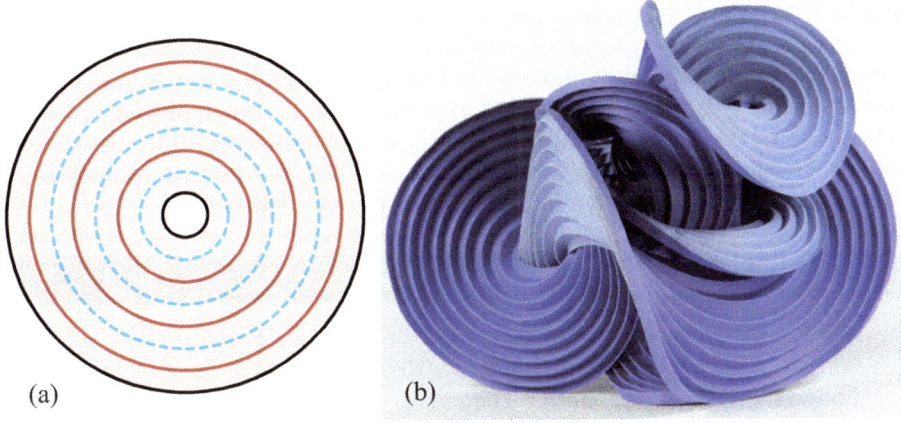

Figure 8.8 (a) Annulus formed of concentric-circles crease pattern. (b) Several circular-pleat crease patterns intertwined. [Reprinted by permission of the artists, Erik and Martin Demaine.]

along these diagonals are close to 180°—nearly flat—and not easily perceptible in a paper model.

The proof of nonexistence employs tools from differential geometry, again well beyond what can be presented here. However, there is one aspect of the proof we can explain.

The central square of the nested-squares crease pattern shown in Figure 8.7(c) is effectively a plus-sign crease pattern. The only difference between this pattern and that shown earlier in Figure 5.13 is that the panels are right triangles—half squares—rather than squares. But still the Plus-Sign Lemma 5.1 holds. (Here we are skipping over a differential geometry argument that justifies treating the panels as rigid throughout any folding.) Recall the lemma states that opposite dihedral angles are equal, and only one pair is nonextreme. The lemma means that two of the creases incident to the central point must not be folded and instead remain flat, contradicting the crease pattern in Figure 8.7(c). It turns out that removing one diagonal of the central square, or removing the central square entirely, creating a hole (analogous to the annulus hole in Figure 8.8(a)), still leaves no rigid folding.

Finally, it has been conjectured that, in contrast to the nested-squares pleat, the circular pleat in Figure 8.8(a) does "exist"—it folds precisely along the circular creases without needing additional creases. It demonstrates how difficult is the mathematics behind curved creases that this conjecture is not yet settled, despite strong numerical evidence supporting a positive answer.

> **Open Problem 8.1 Circular Pleat**
>
> Does the circular pleat in Figure 8.8(a) fold without additional creases?

8.4 Curvature

As just illustrated, there remains much that is unknown concerning the mathematics of curved creases, and that which is known relies on differential geometry. In this section we describe three of the most general results obtained to date, without offering even proof sketches. First we need to understand the curvature of a smooth curve in 3D, and for that we start in 2D.

> **Box 8.1 Tangent and Normal Vectors**
>
> A smooth curve in the plane has at each point p a **tangent vector** and an orthogonal **normal vector**. (Recall Box 1.2 on normal vectors.) Imagine driving a car around the curve. Then the tangent vector points in the forward direction of travel, and the normal vector, 90° turned from the tangent, points in the direction of the centripetal acceleration felt as the car bends around the curve. See Figure 8.10 for examples.
>
> A smooth curve in 3D also has a tangent and normal vector at each point of the curve, but in addition a **binormal** vector, orthogonal to both tangent and normal vectors.
>
> A **smooth curve** has no "kink" (technically, no **singularity**) at which there is no evident tangent or normal vector. An example of a nonsmooth curve is a heart-shaped curve, e.g., Figure 8.9.

8.4.1 Curvature in 2D

Every point p of a smooth curve in the plane has an associated curvature: numerically high curvature at a point where the curve turns sharply, and low

Figure 8.9 A curve with two singularities along the vertical midline.

8.4. Curvature

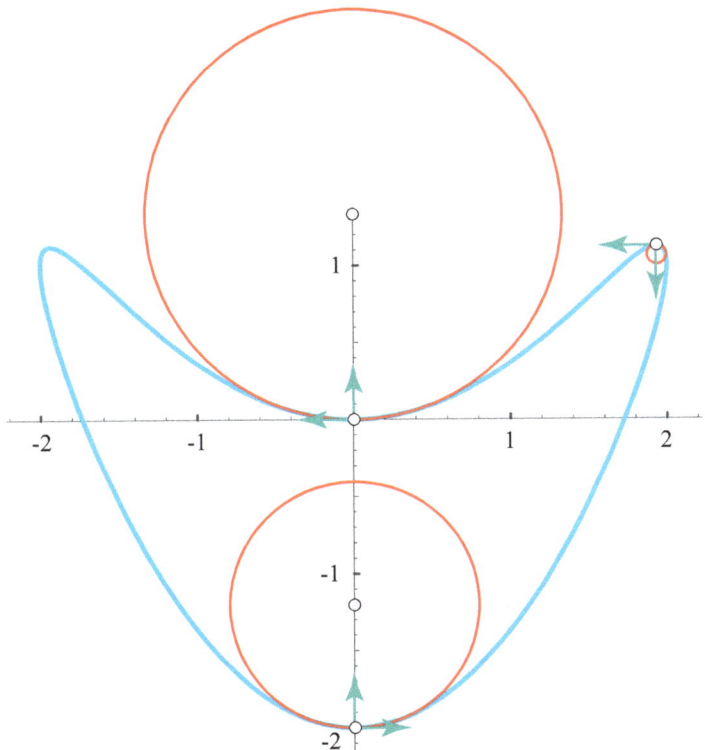

Figure 8.10 Smooth curve (blue) and three osculating circles (red), with tangent and normal vectors (green arrows).

curvature at point on a gentle portion of the curve. Often the symbol κ (Greek kappa) is used to represent the curvature at p.

The curvature κ at a point p is the reciprocal of the radius r of the **osculating** ("kissing") **circle** at p, the "best-fit" circle at p. This relationship is illustrated in Figure 8.10. For example, let $p = (0, -2)$ be the bottom-most point of the blue curve. A calculus calculation shows that the curvature κ at p is $\frac{5}{4}$, and so the radius of the osculating circle is $r = 1/\kappa = \frac{4}{5}$. The osculating circle is tangent to the curve, with its center a distance r along the normal vector to the curve at p. See Box 8.1.

A special case is that a point p along a straight segment of a curve has $\kappa = 0$, and so $r = \infty$. This makes sense if a straight segment through p is viewed as the limit of very large circles tangent at p.

> ### Exercise 8.2 [Understanding] Curvature of Parabola
>
> A calculus calculation shows that the radius of curvature at the vertex of a parabola is twice the parabola's focal length. Assuming this holds, what is the curvature at the vertex of $y = x^2/4$?

> **Box 8.2 Parametric Equations**
>
> Because a curve is 1D, we can imagine it being controlled by a single parameter t determining x and y. So x and y are functions of t: $x(t)$ and $y(t)$, the curve's **parametric equations**.
>
> The (blue) curve in Figure 8.10 has parametric equations
> $$x(t) = 2\sin(t)$$
> $$y(t) = \cos(t) - \cos(2t)$$
> where t ranges from $0°$ to $360°$.
>
> A curve in space is specified similarly. The (blue) curve in Figure 8.11 has these parametric equations (with t ranging from $0°$ to $180°$):
> $$x(t) = \cos(t)$$
> $$y(t) = \sin(2t)$$
> $$z(t) = \cos(t) + \sin(t).$$
>
> Using calculus, the tangent and normal vectors, as well as the curvature κ at each point on the curve, can be calculated as a function of t.

> **Exercise 8.3 [Practice] Circle Parametric Equations**
>
> What are parametric equations for a circle of radius 1? See Box 8.2.

> **Exercise 8.4 [Understanding] Parabola Parametric Equation**
>
> As we've seen in Exercise 8.2, the equation $y = x^2/4$ describes a parabola with vertex at the origin and focus at (0,1). What are parametric equations for this parabola?

8.4.2 Curvature in 3D

The curvature κ at a point p of a smooth curve in 3D mirrors the 2D definition: κ is the reciprocal of the radius of the osculating circle tangent at p. As Figure 8.11 shows, the osculating circle lies in a plane determined by the tangent and normal vectors at p, the **osculating plane**. Although the geometry is more complicated in 3D, the intuition is the same as in 2D: A point on a sharp turn has large curvature and small osculating circle radius, and gently curved portions of the curve have large osculating circle radii. Again straight segments have $\kappa = 0$.

8.5. Three General Properties 131

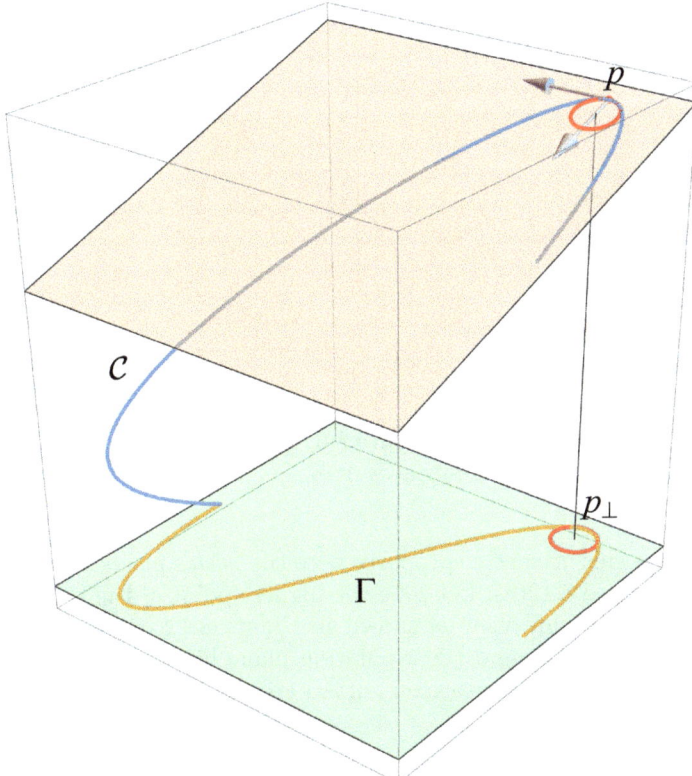

Figure 8.11 Osculating plane (tan) to $\mathcal{C}$ at p. Γ is the 2D projection of $\mathcal{C}$. The osculating circle in 3D has higher curvature than in the 2D projection plane (teal) below.

8.5 Three General Properties

We now turn to the promised general properties of curve creases. The first property was formulated by David Huffman, whose beautiful pinwheel we showed earlier (Figure 8.1).

8.5.1 Osculating Plane Bisection

Theorem 8.1 Osculating Plane Bisection

For any point p on a smooth curved crease $\mathcal{C}$ in 3D, the osculating plane at p bisects the two planes tangent to the surface at p from either side.

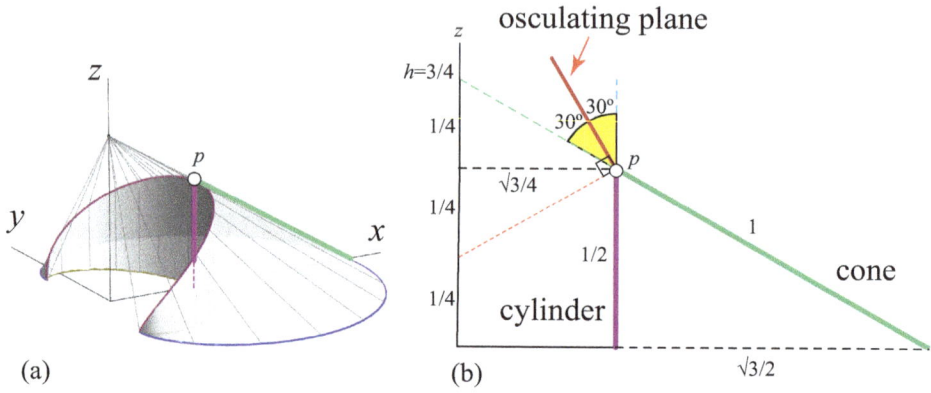

Figure 8.12 (a) Details added to Figure 8.6. (b) *Vesica piscis* side view showing the osculating plane bisecting surfaces at p.

We illustrate this attractive property with the *vesica piscis* shape. We repeat Figure 8.6 as Figure 8.12(a). Let p be the highest point of the curve $\mathcal{C}$, directly above the x-axis. A side view is shown in Figure 8.12(b). The cone and the cylinder intersect at $60°$, and the osculating plane bisects that angle. I believe this bisection property is one reason curved creases are so elegant.

8.5.2 Any Curve: 3D → 2D

The second general property was formulated by Ronald Resch.

> **Theorem 8.2 Any Curve 3D → 2D**
>
> Any smooth space curve $\mathcal{C}$, with nonzero curvature at all points, can be realized as a curved crease Γ in the interior of a piece of paper.

Figure 8.13 shows a smooth space curve on the left, and paper matching the curve a short distance to either side. The reason for the "short distance" caveat is to avoid the paper crossing through itself. The reason for the stipulation that $\mathcal{C}$ has nonzero curvature is that, along a straight segment, although the tangent vector is well defined, the normal vector is not. So the osculating plane is not well defined along such a section, and the claim of the theorem relies on the existence of an osculating plane.

8.5.3 Any Curve: 2D → 3D

Our final general property says that any smooth curve drawn on a flat piece of paper can be creased into 3D.

8.6. Rigidly Foldable Curved Creases

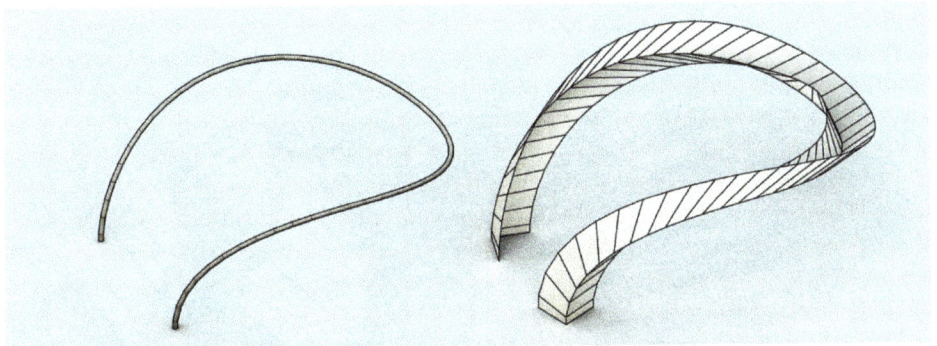

Figure 8.13 Smooth space curve realized by a curved crease. [Reprinted by permission of Tomohiro Tachi.]

> **Theorem 8.3 Any Curve: 2D → 3D**
>
> It is possible to fold an arbitrary smooth curve Γ in the plane into a 3D crease $\mathcal{C}$, with higher curvature at every point.

The higher curvature mentioned in Theorem 8.3 is illustrated in Figure 8.11. There the curvature at p on $\mathcal{C}$ in 3D is 8.5, while the curvature at $p_\perp$ on Γ is 8.0.

Caveats aside, the two theorems (8.2 and 8.3) together show that curved creases are universal in some sense: Any 3D curve can be realized in paper as a curved crease, and any 2D curve drawn on paper can be lifted to a (possibly self-intersecting) curved crease in 3-space.

8.6 Rigidly Foldable Curved Creases

There is a remarkable connection between curved-crease origami and rigidly foldable 1-DOF structures. We sketch two examples, tubes (sketched superficially) and conic creases (in more detail).

8.6.1 Tachi Tubes

Start with an implementation of Theorem 8.2: Given a smooth space curve $\mathcal{C}$ satisfying that theorem, construct a curved folding whose crease lies along $\mathcal{C}$. See Figure 8.14. The surface can be chosen so that the fold angle along the crease is constant, i.e., the same at every point of $\mathcal{C}$. Now form a tube by deriving and arranging three modifications of the original curve, using the constant angle property to ensure that the four surfaces close to a quadrilateral. Underlying these claims are formidable calculations in differential geometry. The end result is an attractive 1-DOF collapsing tube as in Figure 8.14. The inventor,

Figure 8.14 1-DOF Tachi tubes. Note vertices are degree-4. [Reprinted by permission of Tomohiro Tachi.]

Tomohoro Tachi, used these tubes to design a foldable architectural vault composed of tubular arches.

8.6.2 Discretizing Conic Creases

We'll add a bit more detail in our second example, but still the mathematics is too advanced for a full explanation.

In many of David Huffman's designs, the creases follow conics—especially circles, ellipses, and parabolas. For example, his pinwheel model (Figure 8.1) alternates M/V circular arc creases. Another example is his *Arches* tesselation shown in Figure 8.15, which employs parabolic arcs.

Studying conic creases led researchers to investigate when it is possible to **discretize** (polygonize) a conic curve so that it becomes a rigidly foldable 1-DOF construction analogous to Tachi tubes. They discovered that circumscribing a conic curve with tangents at discrete intervals leads to a rigidly foldable design. An impressive example employing several nested parabolic arcs is shown in Figure 8.16.

> **Box 8.3 Parabola Focus at Infinity**
>
> Ellipses and hyperbolas each have two foci, whereas a parabola has just one focus. In the field of projective geometry, the conics are unified by viewing a parabola as having two foci, one "at infinity." This can be intuitively understood by increasing the eccentricity of an ellipse with

8.6. Rigidly Foldable Curved Creases

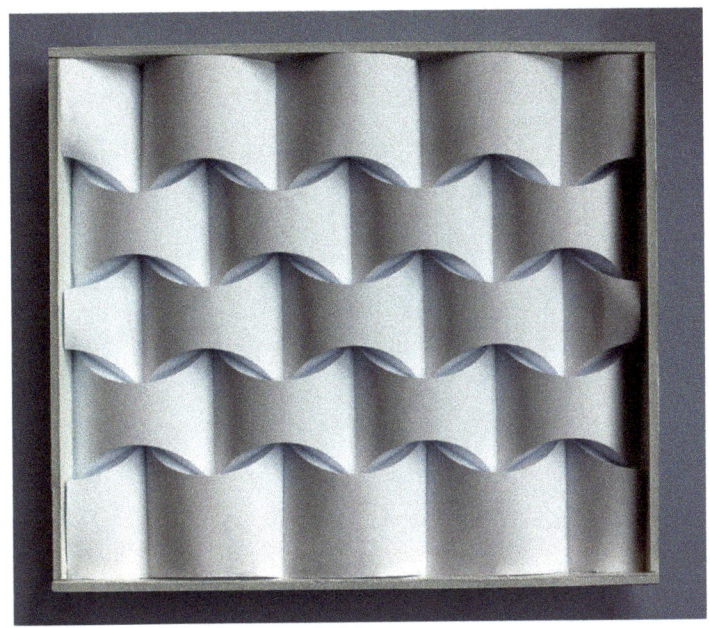

Figure 8.15 David Huffman's *Arches* tesselation. [Used with permission of Taylor & Francis Group, from (Demaine et al. 2011b); permission conveyed through Copyright Clearance Center, Inc.]

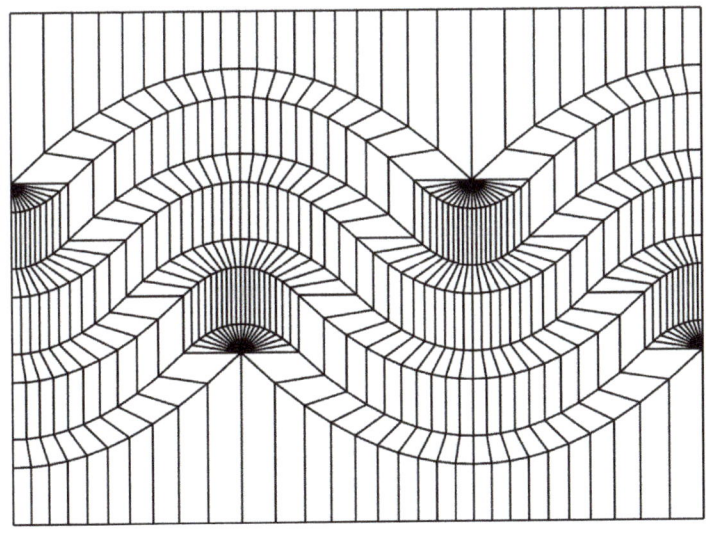

Figure 8.16 Nested discretized parabolas. Figure 10.13 in (Mundilova 2024). [Reprinted by permission of Klara Mundilova.]

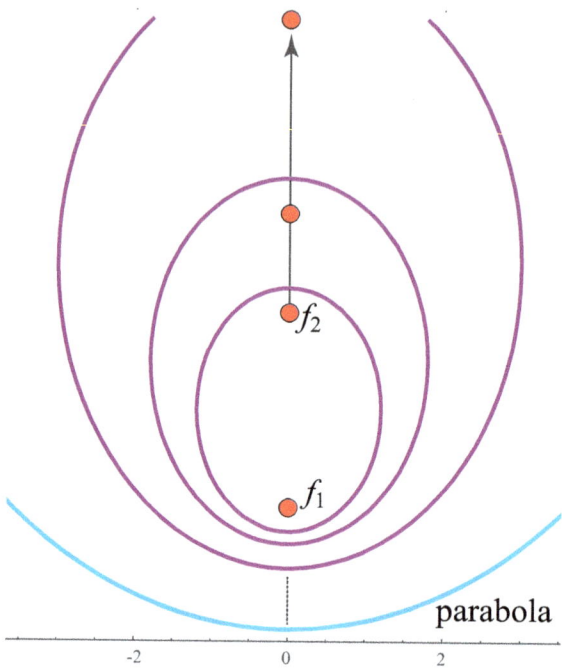

Figure 8.17 A parabola (blue) with focus f_1 is the limit of ellipses whose second focus f_2 approaches infinity.

> one focus f_1 fixed while f_2 separates and heads "toward infinity." In the limit, these ellipses converge to a parabola. See Figure 8.17.
>
> Knowing that a lightray from f_1 of a mirrored ellipse will reflect through f_2 suggests understanding the focus at infinity in terms of a reflecting telescope, which reflects light from infinity (from the focus at infinity) off the parabola to the focus.

8.6.3 Flat-Foldable

Note that every vertex in Figure 8.16 is of degree-4. We will focus on proving that these vertices are (locally) flat-foldable, for then the Degree-4 Folding Theorem 5.2 applies and permits calculation of the 1-DOF dynamics. We first establish generic parabola properties, and then apply these to the pattern in Figure 8.16.

Consider a parabola P with focus at f and focus at infinity f' (Box 8.3). Select two points t_1 and t_2 anywhere along P, and let v be the intersection of the tangents to P at t_1 and t_2. See Figure 8.18.

8.6. Rigidly Foldable Curved Creases

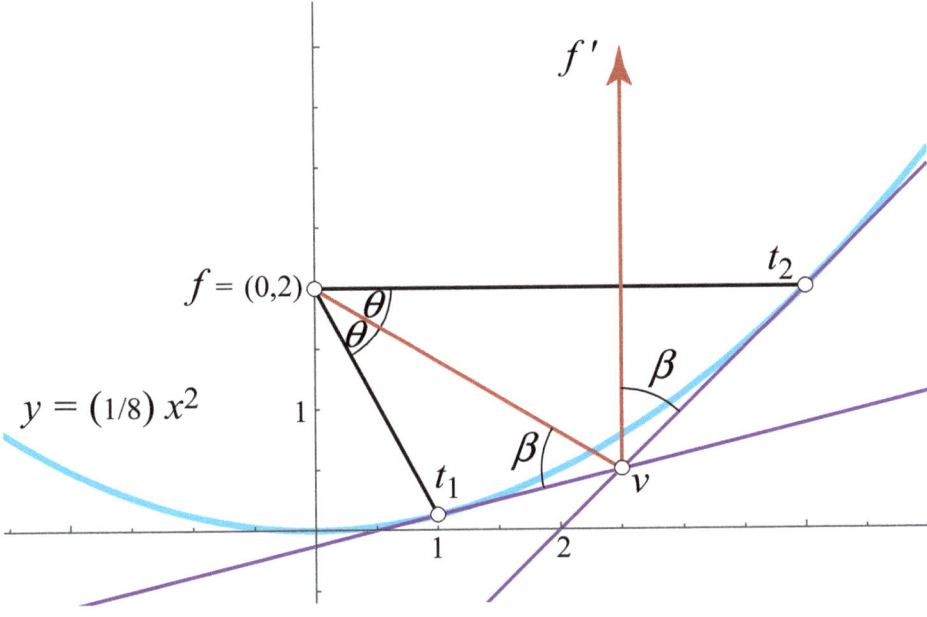

Figure 8.18 $\theta = \angle t_1 fv = \angle t_2 fv \approx 31°$.

We will need properties of parabolas (which also hold for all conics, suitably modified). As the proofs are "elementary" but not straightforward, we opt to leave these as claims. It will be convenient to use the notation $\angle abc$ to mean the measure of the angle at b formed between the segments ab and bc.

> **Lemma 8.1 Parabola Tangents Bisection**
>
> Refer to Figure 8.18. Let t_1 and t_2 be two points on the (blue) parabola, and let v be the intersection of the (purple) tangents at these two points. Then the (red) segment fv, where f is the focus, bisects the angle $\angle t_1 f t_2 = 2\theta$.
>
> Moreover, if f' is the focus at infinity, $\angle fvt_1 = \angle f'vt_2 = \beta$.

> **Exercise 8.5 [Practice] Circle Tangent Bisection**
>
> Prove the bisection property in Lemma 8.1 for a circle rather than a parabola.

The degree-4 crease pattern around a vertex v uses what are called **natural rule lines** for two of the four creases, connecting v to the focus f of the

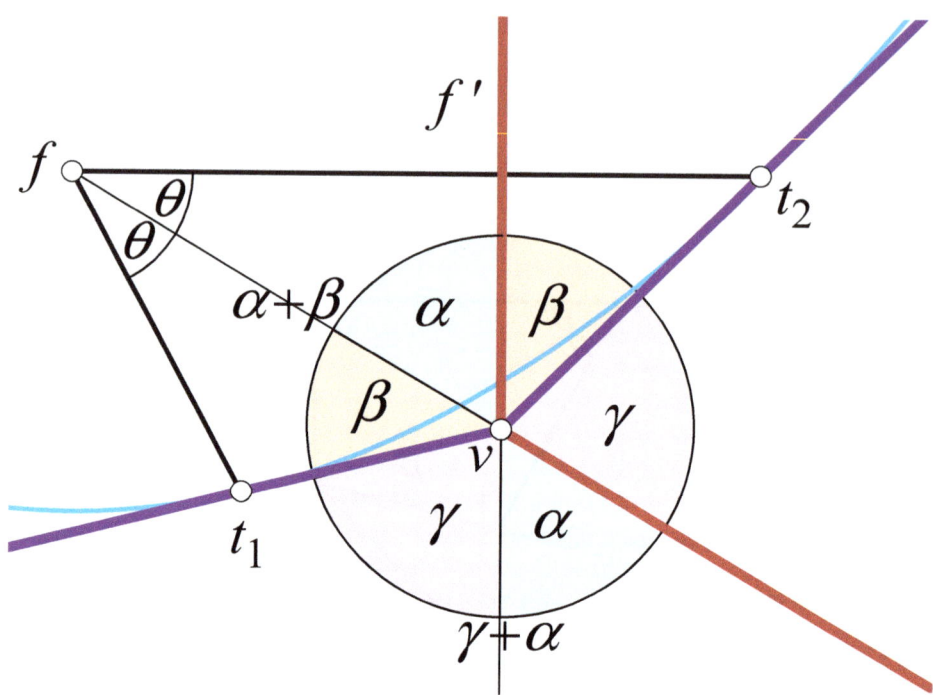

Figure 8.19 Angles around v in Figure 8.18. The red and purple lines are the four creases incident to v.

parabola, and to the focus at infinity f'. The other two creases are the tangents to the parabola. Figure 8.19 shows the creases incident to v.

In order to prove that the degree-4 vertex v is flat-foldable, we need to establish that Kawasaki's Theorem 3.2 holds, i.e., that the alternating sum of the four angles incident to v is 0. Analyzing the six marked angles incident to v in Figure 8.19, we see the following:

- The two α angles are determined by lines through the two foci: the line containing vf and that containing vf'.
- The two β angles are part of the claim of Lemma 8.1.
- That the two γ angles are equal is determinied by the two β angles.

Now we can form the alternating sum of the four incident angles, starting at the eastern γ and proceeding counterclockwise

$$\gamma - \beta + (\alpha + \beta) - (\gamma + \alpha)$$
$$= (\gamma - \gamma) + (-\beta + \beta) + (\alpha - \alpha) = 0°.$$

The cancellation shows that the alternating sum is zero and so Kawasaki's Theorem holds and establishes that v is flat-foldable. M/V assignments could V-fold edge vf' and M-fold the other three edges. Then, by the Degree-4 Folding

8.6. Rigidly Foldable Curved Creases 139

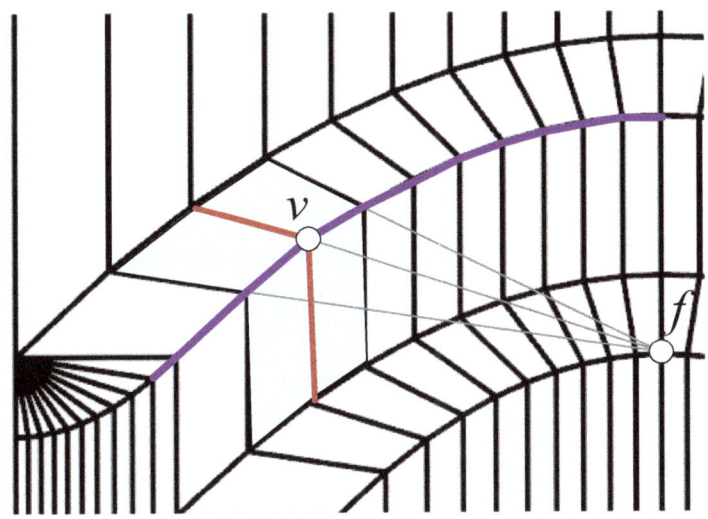

Figure 8.20 Detail from Figure 8.16.

Theorem 5.2, the dihedrals along edges vt_1 and vt_2 are equal while the dihedrals along vf and vf' are equal in magnitude and opposite sign.

We return to Figure 8.16 and highlight where the parabola analysis fits in to that construction. In Figure 8.20, the vertical creases all pass through the focus f' at infinity. The "horizontal" red crease extension passes through the focus f, as do all the creases along the same parabola. The foci for the nested parabolas are vertically stacked under f, rendering them compatibly overlapping one another. This means that the local 1-DOF dynamics extend to global dynamics in a manner similar to the way the Miura map fold dynamics (Section 5.3.6) and the square twist dynamics (Section 5.4.4) are controlled by 1 DOF. The end result is a 1-DOF mechanism that can squeeze the entire construction horizontally and vertically, as illustrated in an analogous Huffman construction in Figure 8.21.

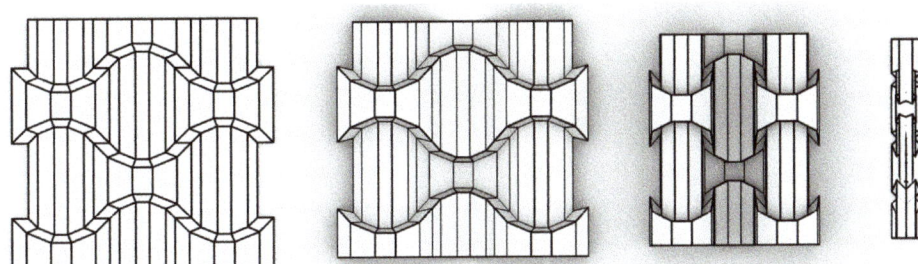

Figure 8.21 1-DOF squeezing and compressing motion of a Huffman design. [Reprinted by permission of Klara Mundilova.]

The beautiful and intricate mathematics of curved creases remains an exciting, active area of research.

8.7 Technical Notes

Sec. 8.1: Introduction Resch (1974). Huffman (1976). Ekaterina Lukasheva (2021). The Demaines: Demaine et al. (2011b). See Demaine et al. (2015) for an overview of curved creases in art and mathematics.

Sec. 8.2: *Vesica Piscis* The analysis of the *vesica piscis* curved folding is due to Mundilova and Wills (2018) and Mundilova (2019).

Sec. 8.3: Pleated Hyperbolic Paraboloid Analysis of the hyperbolic paraboloid: Demaine et al. (2011a), Demaine et al. (2022).

Sec. 8.5: Three General Properties

- Theorem 8.1: Huffman (1976).
- Theorem 8.2: Resch (1974).
- Theorem 8.3: Fuchs and Tabachnikov (1999).

Sec. 8.6: Rigidly Foldable Curved Creases Tachi Tubes: Tachi (2013). Discretizations of conic crease patterns: Demaine et al. (2022), Mundilova (2024).

9

Self-Folding Origami

9.1 Introduction

Recent decades have seen an explosion of interest in *self-folding origami*: rigid origami but where the hinges close or open automatically, initiated by some type of "switch" or "trigger" in the environment. Most aspects of self-folding origami are engineering challenges specific to its myriad applications. Deep mathematics underlies these challenges, but they have not been the focus of the community until recently. We'll describe three narrow mathematical topics: waterbomb tubes (Section 9.2), folding polyhedral containers (Section 9.3), and multiple branches in "configuration space" (Section 9.4). But first, we canvas the engineering aspects superficially.

Mechanisms of Self-Folding The mechanisms to achieve self-folding hinges vary with the scale of applications, which can run from nanometers to meters. In all cases there is some trigger initiated by exposure to a specific stimulus. The mechanisms explored include electrically produced heat on a gel, light hitting polystrene, specific chemical reactions, applied magnetic fields, phase-change material (PCM) switches, springs, or even motors at larger scales.

Applications of Self-Folding We already mentioned in Chapter 5 that solar cells and mirrors have been deployed in outer space, unfurling automatically. One can view this as *self-unfolding*. Self-folding has different aspects and different applications (and, as we'll see, is more difficult to achieve). Applications include drug delivery, self-assembling robots, expanding stents, self-cleaning surfaces, reconfigurable antennas, and microelectromechanical systems (MEMS).

9.2 Waterbomb Tubes

Origami tubular structures have been explored since the 1970s, but were only recently used in self-folding engineering applications. Among the most interesting applications is the innovative use of origami tube designs to serve as

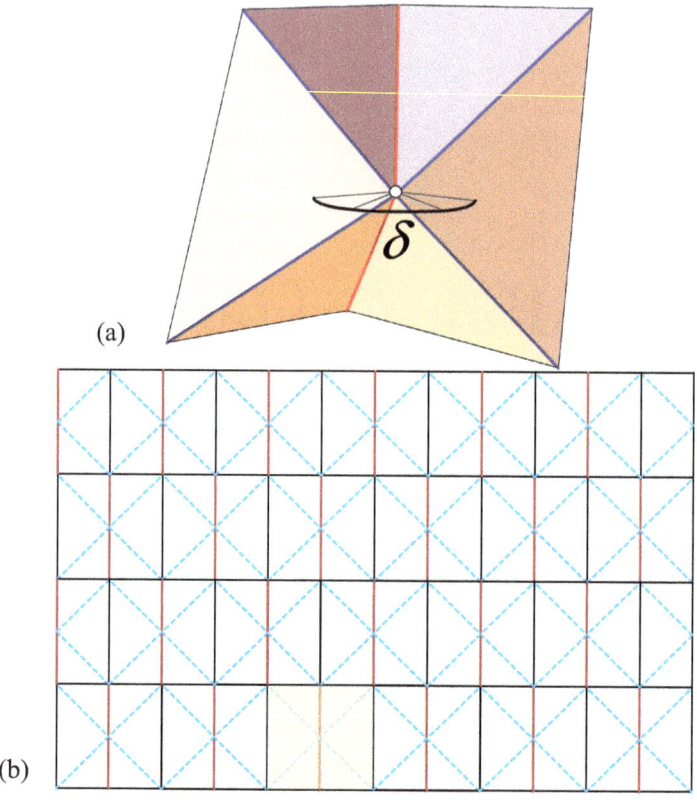

Figure 9.1 (a) Waterbomb base, dihedral angle δ. (b) Waterbomb tube template; one molecule shaded.

medical stents, inserted into a narrowed artery and then expanded, triggered by body heat or warmed via a catheter or via pneumatic pressure. We'll describe one such design based on the waterbomb-base molecule. This design is a type of self-unfolding, but shares with self-folding the characteristic of being triggered to self-morph into a more applicable shape.

Recall from Figure 6.4 that the waterbomb crease pattern forms a degree-6 vertex at the center, with an X-shape of four M-folds crossed by two V-folds. The model is composed of inverted waterbomb molecules—four V-folds and two M-folds, as shown in Figure 9.1(a). The tube is built around a regular polygon; we'll use a regular hexagon. A tube of length n is then built from $6n$ molecules in the 6×4 pattern shown in Figure 9.1(b), with every other row shifted half a unit horizontally. Let δ be the dihedral angle between the two large triangles of a waterbomb molecule, as depicted in Figure 9.1(a).

Joining the left and right edges of the template forms a cylindrical tube. The tube has many degrees of freedom, but if we insist (or arrange) that in

9.2. Waterbomb Tubes

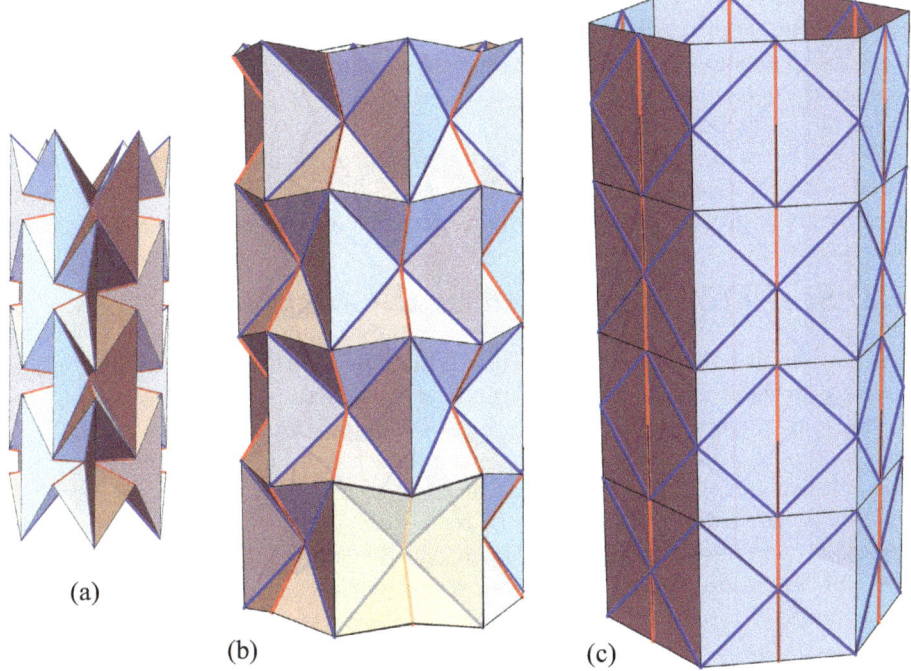

Figure 9.2 Waterbomb tube. (a) $\delta = 66°$. (b) $\delta = 144°$. One molecule shaded. (c) $\delta = 180°$.

each molecule the two M-fold angles are the same, and the four V-fold angles are the same, then the tube has a 1-DOF motion determined by δ. Figure 9.2 shows the tube at three different δ values. In each of the three, all partially folded molecule configurations are geometrically identical. Note that the tube radius increases with δ as does its height—stretching the length expands the radius. This feature of the structure is called a "negative Poisson ratio," useful for the stent application.

> **Exercise 9.1 [Challenge] Asymmetric Waterbomb**
>
> Describe the shape that results when in the waterbomb molecule in Figure 9.1(a), the dihedral angle at the lower M-fold is 180° (so opened flat) while the dihedral along the upper M-fold is 0° (so folded closed).

One can imagine a smooth expansion of the tube from Figure 9.2(a) transitioning to Figure 9.2(c) in concert with increasing δ. However, this is not in fact the true dynamics of the structure. First, the shapes shown are the only ones for which all folded molecules are geometrically identical. At other values of δ, the odd and even rows in Figure 9.1(b) fold differently. For example, at $\delta = 60°$, the large triangles in odd-row molecules collapse to touching. Second, the tube expansion cannot be achieved by rigid origami: instead the triangular faces must deform. Third, the tube shape morphs at $\delta = 120°$ to a pineapple shape closed at both ends!

These surprising dynamics are quite complicated, and only recently understood mathematically. Meanwhile scientists have found many uses of waterbomb tubes, from stents to soft robotics to metamaterials to deformable wheel robots.

9.3 Self-Folding Polyhedral Containers

The main motivation for studying self-folding polyhedra from a flat "net" is that hollow polyhedra can be used to encapsulate biological materials to be delivered for various medical applications. Such containers must be quite small—10 nanometers to millimeters—and large numbers can be manufactured in parallel through lithographic methods, the same technology used for fabricating silicon wafers. A secondary motivation is to better understand how nature self-assembles nanometer viral capsids via the analogy to polyhedral containers. We will set aside the motivation and applications, and concentrate on the mathematics.

> **Box 9.1 Convex Polyhedron**
>
> A ***polyhedron*** is a 3D gem-like shape built from flat faces. The most familiar polyhedra are the five Platonic solids: tetrahedron, octahedron, cube, dodecahedron, isocsahedron. These five polyhedra are ***regular*** in that all faces are the same regular polygon, arranged similarly about its vertices. So the cube has six square faces, three incident to each vertex. The Platonic solids are also ***convex***, in that they have no dents; see Box 6.2 on convexity. Nonconvex polyhedra are important, but difficult to self-fold.

9.3.1 Convex Polyhedron Net

A ***net*** for a convex polyhedron (see Box 9.1) is an unfolding of the surface obtained by cutting edges, unfolding to a single, nonself-overlapping connected planar piece. It is a long-unsolved problem to decide if every convex polyhedron has a net: It could be that every way to cut open some particular convex polyhedron unfolds with overlap. For the purposes of lithographic fabrication, it is crucial that the planar net avoids overlap. Fortunately, all the regular polyhedra have nets, in fact many nets. For example, the cube and the octahedron each have 11 nets: see Figure 9.3.

Several experimental studies have explored the ***yield*** of properly folded polyhedra in large batches of fabricated nets. Polyhedra explored include

9.3. Self-Folding Polyhedral Containers

Figure 9.3 The 11 distinct nets for a cube.

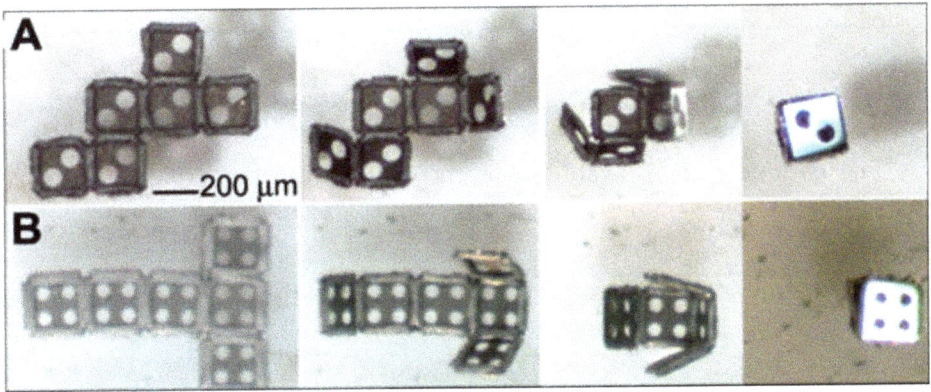

Figure 9.4 Two distinct folding dynamics during self-folding. Cube side length is 200 μm (less than 1/100th of an inch). In Figure 9.3 (L-to-R), (A) tenth net; (b) third net. [From Azam et al. (2009).]

the cube, the octahedron, and the dodecahedron. The cube's 11 nets can be experimentally explored exhaustively. For example, Figure 9.4 shows snapshots of two different 200 μm cube nets successfully self-folding. However, the dodecahedron has a remarkable 43,380 (nonoverlapping) nets, so searches for the "best" nets necessarily must sample from this vast supply.

> **Exercise 9.2 [Understanding] Cut Trees**
>
> A *cut tree* is the graph formed by the collection of edges cut to unfold a polyhedron to a net. Sketch the cut trees that produce the two nets in Figure 9.4.

Experiments show that typically about half of self-foldings have some type of misfolding. One group of researchers found that nets with the lowest "radius of gyration" had the highest yield; see Box 9.2. Another group found that the smallest convex hull perimeter predicted the best yield; see Exercise 9.3. In general, compact nets lead to the highest yields.

> **Box 9.2 Radius of Gyration**
>
> The *radius of gyration* of a polygon is the "root mean square" distance from its centroid to points in the polygon. The root mean square, or RMS, is just what the words say: the square root of the mean of the squares of the distances. Ignoring details, this is a measure of compactness: a small radius of gyration indicates the polygon is compactly arranged around its center of mass.

> **Exercise 9.3 [Practice] Convex Hull Perimeter**
>
> The *convex hull* of a polygon is the shape taken by a rubber band wrapped around the polygon. In Figure 9.5, which net has the smaller perimeter?

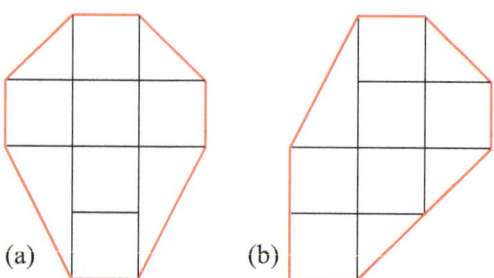

Figure 9.5 Convex hulls of two cube nets.

Although the less than 100 percent yield on self-folding polyhedra experiments is largely due to the formidable engineering challenges, there is in addition an underlying mathematical problem: multiple "folding pathways" in configuration space. We now turn to this topic.

9.4 Paths in Configuration Space

The mathematics behind self-folding relies on the notion of a configuration space. The 3D configuration of a rigid origami model is determined by the dihedral (or equivalently, the fold) angles at each hinge. The **configuration space** is an abstract n-dimensional space, each point of which corresponds to the n hinge angles. Although it is difficult to imagine n-dimensional space when n is more than three, we only need that its points are n real numbers, the n hinge angles. In technical language, n-dimensional space is notated $\mathbb{R}^n$, where $\mathbb{R}$ stands for real numbers (see Box 1.1.)

Any particular folding development of an origami model—for example, a cube net folding, or a Miura map tesselation unfurling—corresponds to a 1D path in the $\mathbb{R}^n$ configuration space. Each point on the path corresponds to the specific n hinge angles of the model configuration at a particular time. The path is a connected series of these snapshots, with all the n hinge angles changing at once.

As we learned in Chapter 5, constraints between the angles reduce the degrees-of-freedom (DOF) below n, sometimes drastically. Both the Miura map and the square twist have just 1 DOF even though n can be large, especially in tesselations. If a model has 1 DOF, then it always follows the same path in the configuration space: It is constrained to follow that one path. A 2-DOF model's dynamics can follow an infinite number of different paths in $\mathbb{R}^n$, but the paths all lie on a 2D surface inside $\mathbb{R}^n$. And this description generalizes for k-DOF models.

When the number k of DOFs is less than n, it is common to represent the full $\mathbb{R}^n$ by $\mathbb{R}^k$, with the understanding that those k DOFs determine the remaining hinge angles. Viewing the dynamics of an origami model as a path in configuration space is a powerful analytical tool.

Perhaps the two most studied configuration spaces are (1) flat-foldable degree-4 vertices, and (2) "trifold" degree-6 vertices. Chapter 5 focused on (1), claiming in the Degree-4 Folding Theorem 5.2 that all these have 1-DOF dynamics. So these models follow a particular 1D path in $\mathbb{R}^4$: One hinge angle determines all four. Now we focus on (2).

9.4.1 Degree-6 Vertex

We concentrated in Chapter 5 on degree-4 vertices because the powerful Theorem 5.2 established specific dynamics for flat-foldable degree-4 vertices. It is a general rigid origami fact that a vertex of degree-k (flat-foldable or not) has $k-3$ DOFs, which we leave a claim without proof. In particular, a degree-4 vertex has 1 DOF, and a degree-6 vertex has 3 DOFs. Many flasher designs (Section 5.2.3) naturally employ degree-6 vertices, but the one we showed in Figure 5.7(a) specifically avoids degree-6 vertices, relying on degree-4 vertices

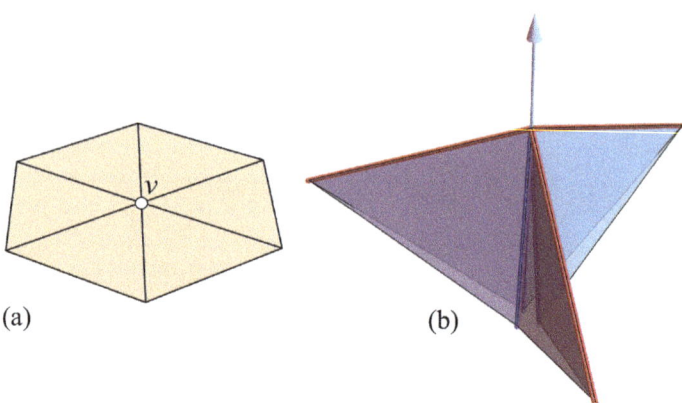

Figure 9.6 (a) Flat hexagon. (b) Degree-6 vertex creases nearly fully folded. Animation: https://cs.smith.edu/~jorourke/MathOrigami/.

to ensure 1-DOF deployment dynamics. All vertices of the waterbomb tube studied in Section 9.2 have degree-6, which is one reason why its dynamics is so difficult to analyze.

Here we focus on an isolated degree-6 vertex v at the center of a flat, regular hexagon formed of six equilateral triangles; see Figure 9.6(a). A little exploration with a creased paper hexagon verifies that this structure has 3 DOFs: fixing the hinge angles of every other edge incident to v determines the intermediate angles. We now analyze a 1-DOF version of this flat hexagon.

Let the edges be labeled $e_1, e_2, \ldots, e_6$ and the hinge fold angles $\rho_1, \rho_2, \ldots, \rho_6$. Here we use fold angles rather than dihedrals to match the literature, so all the angles ρ_i are 0° when the hexagon is flat. Positive ρ_i means a mountain fold, negative a valley fold. We will study **trifolds**, which artificially constrain the folds at every other edge to be identical. Then the vertex v has just 1 DOF. Figure 9.6(b) shows the configuration when the three fold angles are each nearly 180° M-folds, and so the intermediate angles are nearly flat 0° V-folds. The M/V pattern resembles the rabbit-ear molecule (Figure 6.6) prior to uniaxial alignment.

> **Exercise 9.4 [Understanding.] Pinched Hexagon**
>
> The M/V pattern in Figure 9.6(b) alternates M and V: MVMVMV.
>
> (a) Is it possible to fold the six creases MMVMMV and achieve a flat state?
>
> (b) If again the six hexagon creases are MMVMMV, but the six hinges and triangles are rigid, what shape results from the flat state if all the M hinge fold angles are increased at the same rate?

9.4. Paths in Configuration Space

Now we finally connect to self-folding. Suppose we'd like to self-fold from the flat hexagon state to the "trident" shape in Figure 9.6(b), perhaps as one component in a larger dynamic origami construction/structure. The first hurdle is that the model can pop-down as equally well pop-up: the two are symmetric possibilities. This could be controlled if necessary by biasing the hinge springs (or whatever mechanism is employed) to M-fold along $\{e_1,e_3,e_5\}$ forcing V-folds along $\{e_2,e_4,e_6\}$. But there is a surprise: There are two different configurations, both popped-up and both with the same M/V pattern! These configurations are not easy to feel when manipulating a paper model by hand, but they exist mathematically. This may not matter if the application only requires the final state in Figure 9.6(b). But it could be that the path in configuration space needs to follow one configuration path over another, say, to avoid collisions in folding polyhedral nets.

A representation of the configuration space for the hexagon is shown in Figure 9.7. Although under the trifold assumption the model has 1 DOF, and so could be represented as a 1D configuration space, it is more revealing to

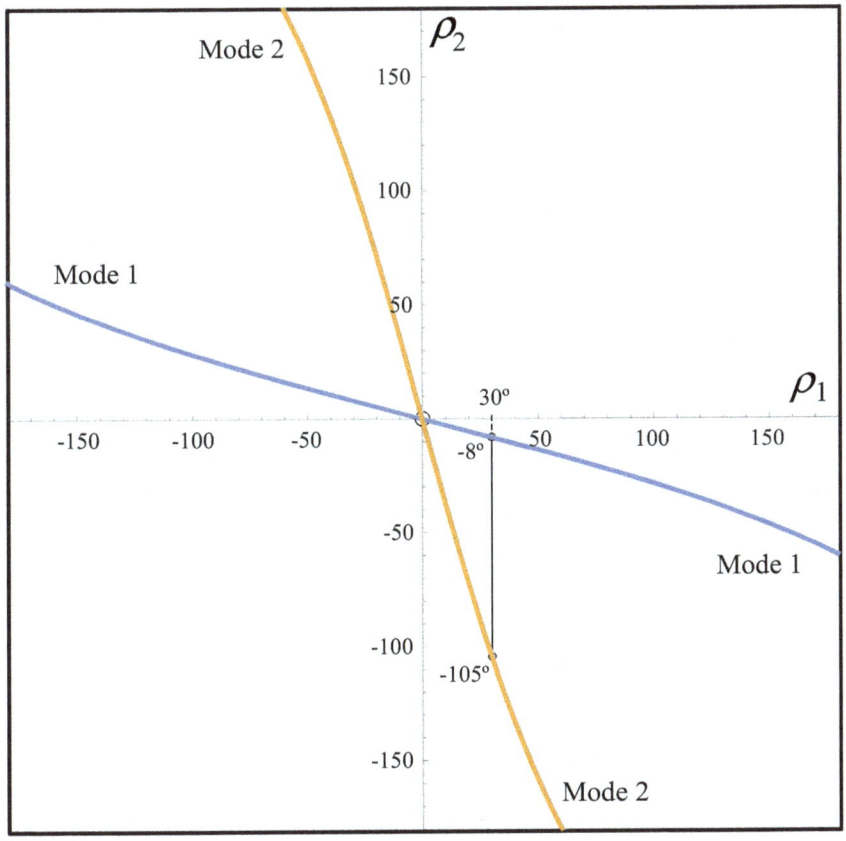

Figure 9.7 Two modes in $\rho_1 - \rho_2$ configuration space.

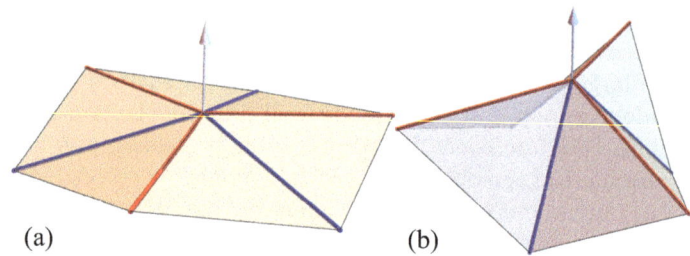

Figure 9.8 $\rho_1 = 30°$. (a) Mode 1. (b) Mode 2. Mode 1 animation (same as in Figure 9.6): https://cs.smith.edu/~jorourke/MathOrigami/.

represent the dynamics in a 2D space where the two dimensions represent the fold angles $\{\rho_1, \rho_2\}$ (which, under the trifold assumption, determine all six angles). There are two 1-DOF paths in the space, crossing at the flat state origin $\rho_1 = \rho_2 = 0$. A challenging calculation shows that the equations for these two paths ("modes") are remarkably similar to the half-tangent equations in Section 5.3.5, Equation (5.1)—here quarter-tangent equations—hinting at deep connections not yet fully understood:

$$\tan(\rho_1/4) = \mu \tan(\rho_2/4)$$
$$\tan(\rho_2/4) = \mu \tan(\rho_1/4)$$

where $\mu = -(2+\sqrt{3})$. One can see that the curve for Mode 2 is a reflection of the Mode 1 curve across the diagonal $\rho_2 = -\rho_1$. Figure 9.8 shows the two configurations reached when $\rho_1 = 30°$. Note that the Mode 2 path folds the model more quickly with respect to ρ_1 than does the Mode 1 path: At $\rho_1 = 30°$, the Mode 1 path has only folded $|\rho_2| = 8°$ but the Mode 2 path has folded ρ_2 more than $100°$.

> **Exercise 9.5 [Understanding] Plus-Sign Configuration Space**
>
> Sketch out the pathway(s) in $\alpha - \beta$ configuration space, where α and β are fold angles, for the plus-sign described in Section 5.3.1 (Figure 5.13).

9.4.2 Distractor Paths

Without the trifold assumption, the configuration space requires three dimensions to represent $\{\rho_1, \rho_3, \rho_5\}$. Imagine then an infinite number of "modes," each a path in this configuration space, forming a kind of "spaghetti" of paths, each cinched to passing through the flat-state origin $\rho_i = 0$. Then, starting from this

9.4. Paths in Configuration Space

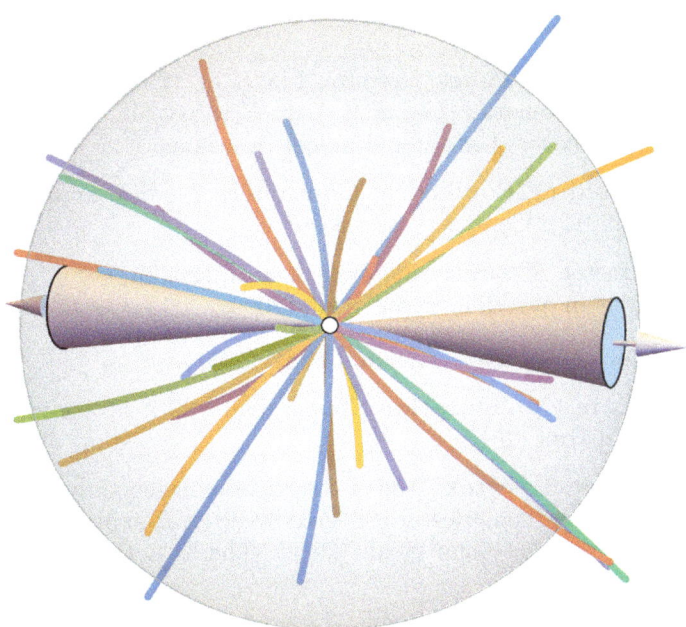

Figure 9.9 Abstract configuration space. Desired path must move from the flat-state center point (white) into narrow ±cones, avoiding distractor paths (multicolored).

flat state, there are numerous paths available to the model. Such paths have been called ***distractor paths*** because they can "distract" the origami model from following the desired path through configuration space. See Figure 9.9. This partly explains the less-than 100 percent yield we saw in folding polyhedral nets (Section 9.3.1).

Even with the well-behaved Miura map, there are an exponential number of distractor paths to avoid between Miura units. So even though a Miura tesselation *unfolds* with controlled regularity, one could not rely on the activation of a single 1-DOF Miura unit to control the *folding* of an entire tesselation. Instead multiple hinge actuators would be necessary. In this sense, folding from the flat state is more difficult than unfolding from the folded, compressed state. This is intuitively evident when struggling to refold a physical map which was easy to unfold!

The promise of self-folding in many application areas makes this a highly active research area. As one researcher put it,

> "The flurry of recent activity has quickly yielded a number of tools for designing self-folding structures. Yet, we have barely scratched the surface of what is possible."

9.5 Technical Notes

Sec. 9.2: Waterbomb Tubes I rely on Ma et al. (2020) for understanding the complex kinematics. Use as a stent: Kuribayashi et al. (2006). Lang (2017, Ch. 2) covers history and model variations.

Sec. 9.3: Self-Folding Polyhedral Containers Medical applications: Randall et al. (2012). Compactness measures:

- Perimeter of convex hull: Jungck et al. (2022).
- Radius of gyration: Azam et al. (2009).

Dodecahedron nets: Horiyama and Shoji (2011).

Sec. 9.4: Paths in Configuration Space Figure 9.7 is based on Fig. 8 in Tachi and Hull (2017).

Sec. 9.4.1: Degree-6 Vertex Degree-6 vertices in flashers: Lang et al. (2016). The complex mathematics of a degree-6 vertex is explored in Tachi and Hull (2017) and Farnham et al. (2022). The $k-3$ DOFs claim is Thm. 13.20 in Hull (2020).

Sec. 9.4.2: Distractor Paths Figure 9.9 is based on Fig. 1(c) in Stern et al. (2017), which contains a thorough discussion of distractor paths. The final quote is from Santangelo (2017).

10

ORIGAMIZER

10.1 Introduction

This final chapter addresses the "ultimate challenge in computational origami design": How to fold any shape desired. This of course depends on what constitutes a "shape." One line of investigation aims for any polyhedron. We've encountered polyhedra before, but usually convex polyhedra—those without dents (Box 9.1). Nonconvex polyhedra can be quite complicated, for example, the "Stanford Bunny" shown in Figure 10.1 is a nonconvex polyhedron. This particular model was developed by Stanford researchers in 1994 as a test bed model for computer graphics algorithms, and has since become an iconic geometric shape. The notion of "polyhedron" can be broadened to include flat, two-sided surfaces, and *polyhedral manifolds*, shapes with one or more holes, for example, a torus. What is excluded here is curved creases, but as we've seen in Chapter 8, often curved creases can be discretized, resulting in polyhedra.

> **Box 10.1 Topology**
>
> The emphasis in this book has been on geometry, which focuses on metric distances and angles. The field of *topology* describes surfaces that can be deformed but not sliced. The standard example is that a coffee cup is *topologically equivalent* to a donut because they both have one hole: Each is a deformed torus. The 2D piece of paper being folded is a *topological disk*: It has one boundary "loop" or *cycle*. The Stanford Bunny is a topological disk with boundary the rim of the base.

To "origami-fold" a polyhedron $\mathcal{P}$ we take to mean creating the faceted surface of $\mathcal{P}$ by creasing and folding a single piece of paper. In the next section we describe a "strip algorithm" which in some sense meets the challenge, but only by "cheating"—the final folding of $\mathcal{P}$ is quite unsatisfactory. Nevertheless, the strip algorithm employs interesting mathematics (and origami folds), and

Figure 10.1 The original model has 35,947 vertices, but here we show a version with 2503 vertices. [From O'Rourke (2022). Reprinted by permission of Cambridge University Press. Model from the Stanford 3D Scanning Repository.]

sharpens our sense of what *is* a satisfactory folding of $\mathcal{P}$. We then conclude in Section 10.3 with a description of the (amazing!) ORIGAMIZER algorithm.

10.2 Strip Algorithm

The strip algorithm is perhaps best viewed as wrapping a physical polyhedron with one long strip of paper. We'll describe it in several steps, staying short of formal proofs in order to convey the main ideas. The first two steps "prepare" the polyhedron.

(1) Triangulate the polyhedron surface. Partition every face of the polyhedron into triangles. This often introduces coplanar triangles, triangles lying in the same plane. For example, a triangulation of a cube partitions each of its six faces into two coplanar triangles each.

10.2. Strip Algorithm

(2) Refine the triangulation to a Hamiltonian triangulation. A *Hamiltonian triangulation* is one on which you could walk from triangle to adjacent triangle (sharing an edge), stepping on each triangle exactly once; see Box 10.2.

> **Box 10.2 Hamiltonian Triangulation**
>
> Named after William Rowan Hamilton, an Irish scientist who in 1857 created a game that asked for a path stepping on each triangle face of an icosahedron exactly once and returning to the starting triangle. The *icosahedron* is the fifth of the five Platonic solids, composed of 20 equilateral triangles. This is called a *Hamiltonian cycle*, in contrast to a *Hamiltonian path*, which need not close to a cycle, but still must touch each triangle face exactly once. Often it is more convenient to think of Hamilton's game on the edges of a *dodecahedron*, the fourth Platonic solid, whose structure is "dual" to that of an icosahedron.

> **Exercise 10.1 [Challenge] Latin Cross**
>
> Using the unfolding of the cube to the "Latin cross" net (the first net in Figure 9.3), triangulate its six faces so that it is Hamiltonian.

Not every triangulation is Hamiltonian, but there is always a "refinement" that is, partitioning each triangle into several subtriangles. See Exercises 10.1, 10.2, and 10.3.

Now the polyhedron is prepared for wrapping.

> **Exercise 10.2 [Understanding.] Dodecahedron Cycle**
>
> Find a Hamiltonian cycle following the edges of the *Schlegel diagram* of the dodecahedron shown in Figure 10.2.

> **Exercise 10.3 [Challenge] Non-Hamiltonian Refinement**
>
> Refine the non-Hamiltonian triangulation in Figure 10.3 by partitioning triangles with chords so that it supports a Hamiltonian path.

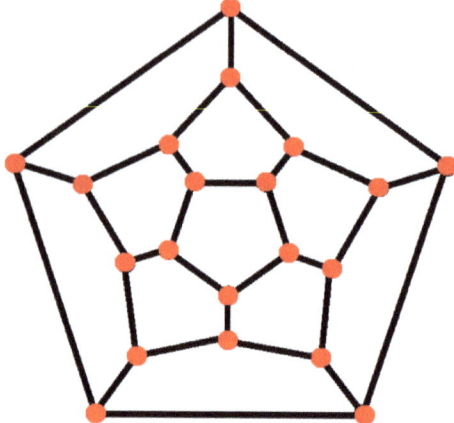

Figure 10.2 A Schlegel diagram of the dodecahedron. [Wikipedia Commons, author Tom Ruen. CC BY-SA 3.0.]

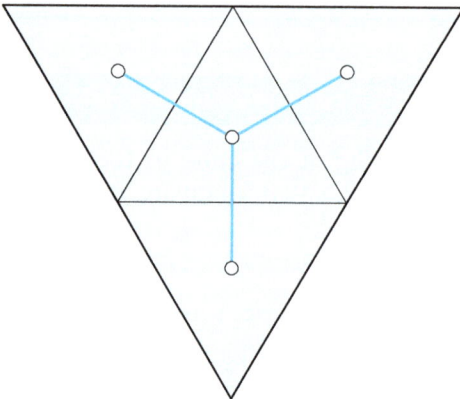

Figure 10.3 A non-Hamiltonian triangle. Its ***dual graph*** is shown, connecting triangles that share an edge.

(1) Fold paper into a long thin strip. Accordion fold the paper, alternating parallel mountain and valley creases. This is among the main senses in which this algorithm produces unsatisfactory folded models.

(2) Zig-zag cover each triangle face, folding over overhang paper ("flash") as needed. See Figure 10.4.

(3) Walk from triangle to triangle along the Hamiltonian path through the triangles, covering each triangle.

(4) Turn between triangles as needed via ***turn gadgets***. A typical turn gadget is shown in Figure 10.5. The goal (Figure 10.5(a)) is to turn a

10.2. Strip Algorithm

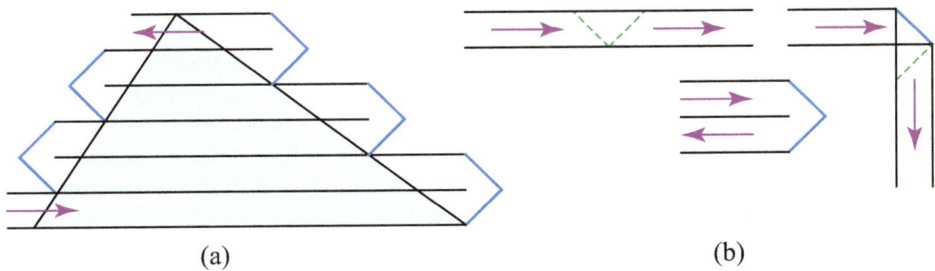

Figure 10.4 (a) Zig-zag covering of a single triangle. (b) How to turn the strip.

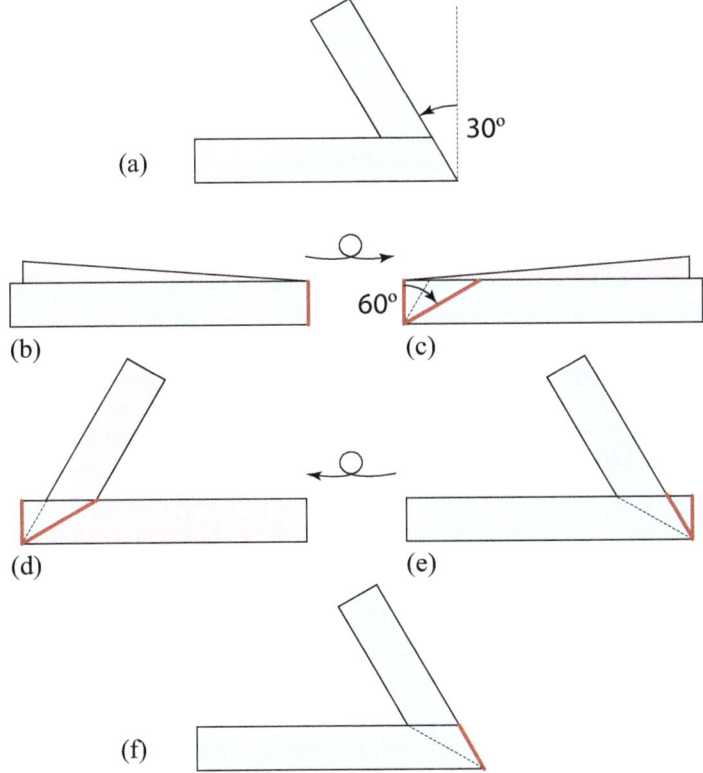

Figure 10.5 (a) The goal. (b,c,d,e) Steps in construction. Tucking the small overhang achieves the goal (f).

horizontal strip 120° from the horizontal, i.e., 30° beyond the vertical. Here we pay attention to the two sides of the paper, blue on the front, pink on the back. Crease the strip in half (Figure 10.5(b)), and flip over (Figure 10.5(c)). Here the desired 30° is shown dashed, and a mountain crease is marked at twice the angle, so 60°. Folding this crease (Figure 10.5(d)) aligns the left edge of the now angled strip with the desired 30° angle. Flip over again (Figure 10.5(e)), now looking at the front again. Finally, mountain-fold the small overhang triangle to hide it behind (Figure 10.5(f)), and we have achieved the goal of Figure 10.5(a).

Even sharper turns may need more than one fold to hide the overhang. If the desired turn from the horizontal is less than 90°, a slightly different gadget may be employed.

Hopefully it is convincing without further details that this algorithm works for any polyhedron. But one can see that, even covering a cube from a square piece of paper would be hugely wasteful of paper, as well as producing a flimsy construction with no structural integrity. The waterbomb cube in Figure 6.4(d) is much superior! However, the algorithm does achieve "folding any polyhedron" in a loose sense. We now turn to ORIGAMIZER, which also achieves folding any polyhedron, but without the strip algorithm's weaknesses.

10.3 ORIGAMIZER Algorithm

To justify calling the ORIGAMIZER algorithm "amazing," see Figures 10.6 and 10.7. It took Tomohiro Tachi 10 hours to fold the model! We will return to details of this stunning example throughout this section.

10.3.1 Weaknesses Overcome

There are three main aspects in which the weaknesses of the strip algorithm are resolved by ORIGAMIZER:

(1) Paper usage is much improved, with ORIGAMIZER often needing only about two to five times the surface area of the model, whereas the strip algorithm might need 1000 times the area. The Stanford Bunny folding in Figure 10.7 uses less than six times the area.

(2) Each triangle face of the strip algorithm's construction includes many *seams*, either creases or subsegments of the paper boundary. This is evident in the zig-zag covering in Figure 10.4. In the final construction by the ORIGAMIZER algorithm, every (convex) face F is seamless: Although the edges of F are seams, the interior of each F is without seams. The only restriction here is that the faces of $\mathcal{P}$ need to be convex. This is easily satisfied by partitioning any nonconvex face into coplanar convex pieces. Thus, for example, ORIGAMIZER applied to a cube will leave all six squares uncreased, in contrast to the waterbomb cube (Figure 6.4(d)). All the faces of the Stanford Bunny in Figure 10.6 are triangles, and you can see in the crease pattern that each is seamless.

(3) Perhaps the most important difference between the constructions of ORIGAMIZER and the strip algorithm is that the ORIGAMIZER foldings are *watertight*. To explain this, we first return to topology (Box 10.1).

10.3. ORIGAMIZER Algorithm

Figure 10.6 ORIGAMIZER Stanford Bunny composed of 374 triangles. [Reprinted by permission of Erik Demaine and Tomohiro Tachi.]

A piece of paper P is a topological disk: it has a single boundary B, a topological circle. So any folding of P must also be a topological disk. The bunny is a topological disk if the base is excluded (as it usually is), so the rim of the base is the topological circle B. To turn a closed polyhedron into a topological disk requires slitting the model somewhere, so that the two sides of the slit together form a topological circle.[1]

Now we return to watertightness. Let the polyhedron $\mathcal{P}$ have boundary B. A watertight folding of $\mathcal{P}$ has no holes, gaps, or slits internal to B. The boundary of the paper can be chosen to be as close as desired to the boundary B of the polyhedron, and all the surface of $\mathcal{P}$ is covered

---

[1] More complicated slitting (cutting a tree) achieves the same goal.

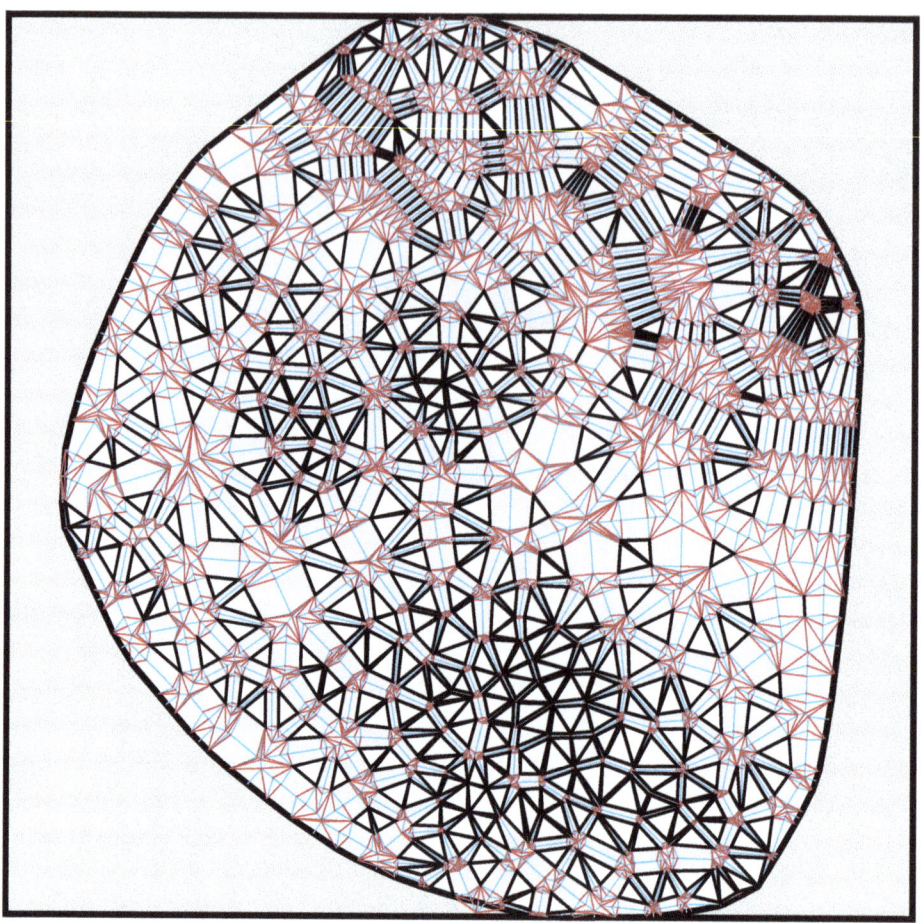

Figure 10.7 ORIGAMIZER crease pattern for Figure 10.7. All (white) triangle faces are seamless. Thick black edges are edges of $\mathcal{P}$. Red edges: M; blue edges: V. [Reprinted by permission of Erik Demaine and Tomohiro Tachi.]

by paper interior to the paper boundary. If the bunny is turned upside down, it could be filled with and retain water.

Figure 10.8 illustrates the difference between the strip algorithm (left)[2] and the ORIGAMIZER algorithm (right) on a saddle surface made of convex quadrilateral faces. The paper boundary in the strip output criss-crosses the interior of the model, whereas ORIGAMIZER places the paper boundary close to the polyhedron boundary.

---

[2] In this example, the strip algorithm presented in Section 10.2 did not need to triangulate.

10.3. ORIGAMIZER Algorithm

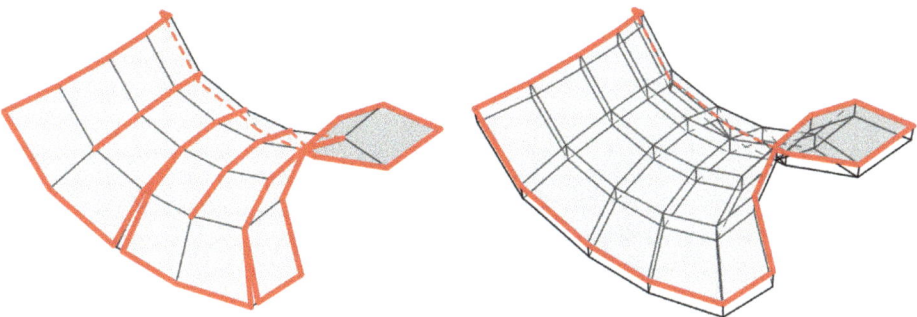

Figure 10.8 Paper boundary is red. Strip algorithm left, ORIGAMIZER right. [Reprinted by permission of Erik Demaine and Tomohiro Tachi.]

10.3.2 The ORIGAMIZER Algorithm

First we should make clear that, as of this writing there are two ORIGAMIZERS. The first is the implemented algorithm, freely available for noncommercial use. In relatively rare circumstances this algorithm fails to work, and in fact did not work on the original Stanford Bunny. However, by extending the boundary of the base by cutting edges to the ears (visible in Figure 10.6), the algorithm worked and produced the crease pattern in Figure 10.7.

The second ORIGAMIZER algorithm is more complicated but proven to work for all polyhedra, under a very broad definition of "polyhedron."[3]

> **Theorem 10.1 ORIGAMIZER**
>
> The ORIGAMIZER algorithm folds a piece of paper to a model of any given polyhedron $\mathcal{P}$, with the model guaranteed to be both seamless and watertight.

Several of the steps in the proof are not present in the implemented algorithm. As our description will be superficial, the distinction between the two versions will not matter: We'll call both ORIGAMIZER.

After this long build-up, the reader may be anxious to learn how the algorithm achieves its amazing results. Unfortunately, it employs advanced mathematics and is so complicated that only a coarse description is feasible.

Topological Disk We illustrate ORIGAMIZER applied to a cube, and to the Standford Bunny—two extremes of possible shapes—very simple and very complex. Because the cube is a closed shape, it needs to be cut so that it becomes a topological disk. Here we use the 3-edge cut v_2, v_3, v_4, v_8: See Figure 10.9, which also includes labeling to help interpret the crease pattern in Figure 10.10. Note

---

[3] Any "oriented polyhedral manifold."

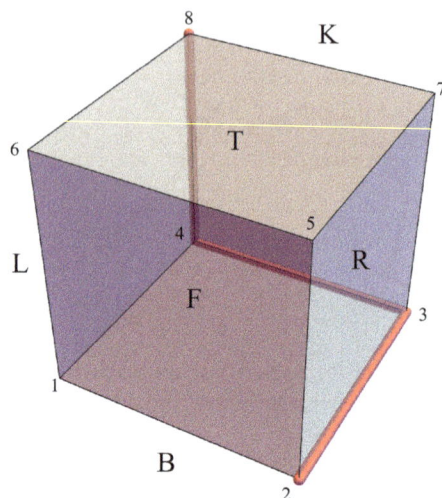

Figure 10.9 Cube faces F, T, R, L, B, K: Front, Top, Right, Left, Bottom, bacK. Vertex indices $1,\ldots,8$. The cut path (red) is v_2, v_3, v_4, v_8.

that because each cut edge has two sides, the boundary B of the cube crease pattern in Figure 10.10 is six edges long.

Mapping Faces to Paper The next step is to map the surface of $\mathcal{P}$ to the plane containing the paper in such a manner that (a) each face in the plane is congruent to its 3D counterpart on $\mathcal{P}$, and (b) no two faces overlap. So the faces are spread out on the paper. One can clearly see the six square cube faces dispersed in Figure 10.10, and perhaps less clearly see the 374 triangle faces in the bunny crease pattern, Figure 10.7. We will not detail this step further, except to say that it employs "nonconvex optimization."

In some sense, the remaining task is now to fold the "extra" paper to draw together the triangles that are adjacent on $\mathcal{P}$. The folded extra paper will be stored on the back side the model, the ***tuck side***, leaving the clean side tiled with seamless triangles.

Edge-tucking Consider drawing together two edges e_1, e_2 shared by triangles F_1 and F_2. A single crease suffices: see Figure 10.11. M-folding e_1, e_2 while V-folding their bisector (Box 7.1) hides the extra paper on the tuck side. Note, however, that the material on the tuck side might collide with other portions of the construction. This can be handled by accordion/pleat folding along the path connecting the two edges, reducing the height on the tuck side.

In the cube crease pattern (Figure 10.10), the edge folds bringing together shared edges, or an edge to a boundary edge, are shaded green. For example, edge $v_7 v_8$ on face T connects to edge $v_7 v_8$ on face K via a (green) fan of five creases, MVMVM. A similar fan connects the two copies of edge $v_3 v_7$ from face

10.3. ORIGAMIZER Algorithm

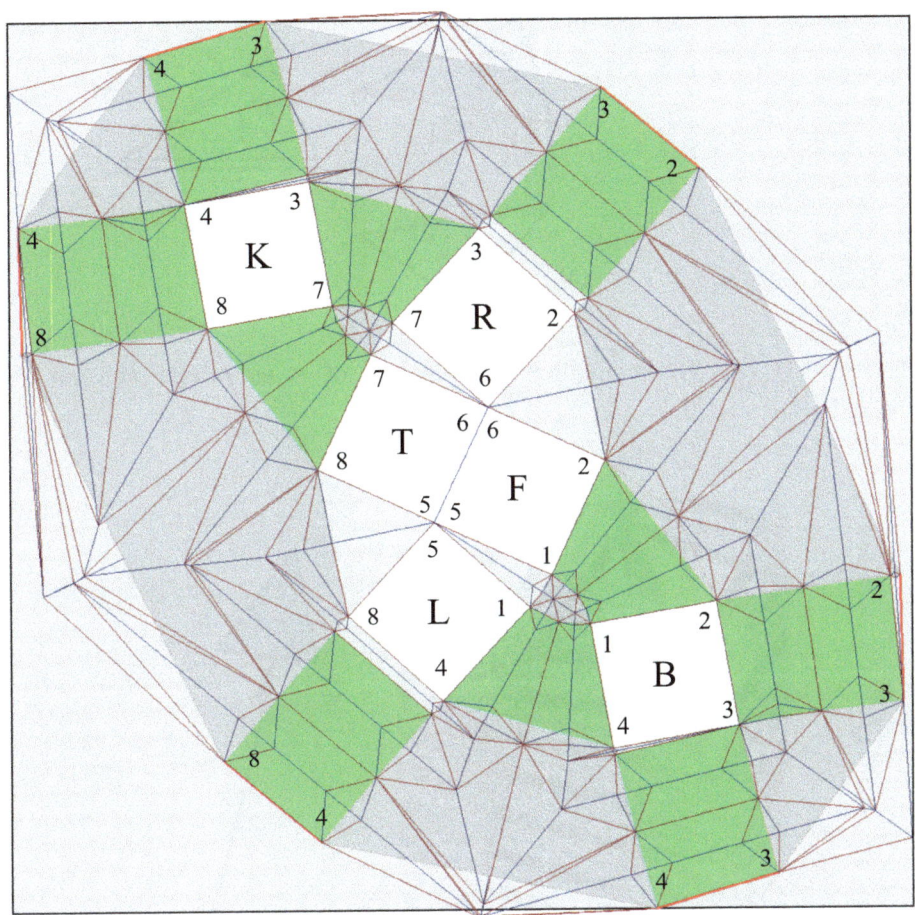

Figure 10.10 ORIGAMIZER crease pattern for cube. Faces white. Edge accordion pleats green. M: red; V: blue. Labels are defined in Figure 10.9. [Reprinted by permission of Tomohiro Tachi.]

R to K. A long-distance connection from edge v_3v_4 of face K to the other copy of this edge on face B crosses the entire crease pattern.

An edge-tucking example drawn from the Stanford Bunny crease pattern (Figure 10.7) is shown in Figure 10.12. The flatness of the bunny's ears requires extensive accordion folding.

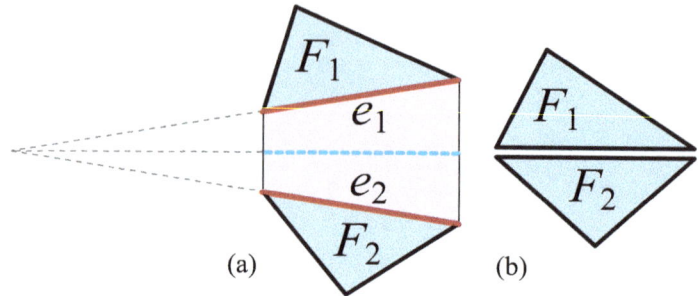

Figure 10.11 (a) Valley folding the bisector of e_1, e_2 brings F_1 and F_2 together (b).

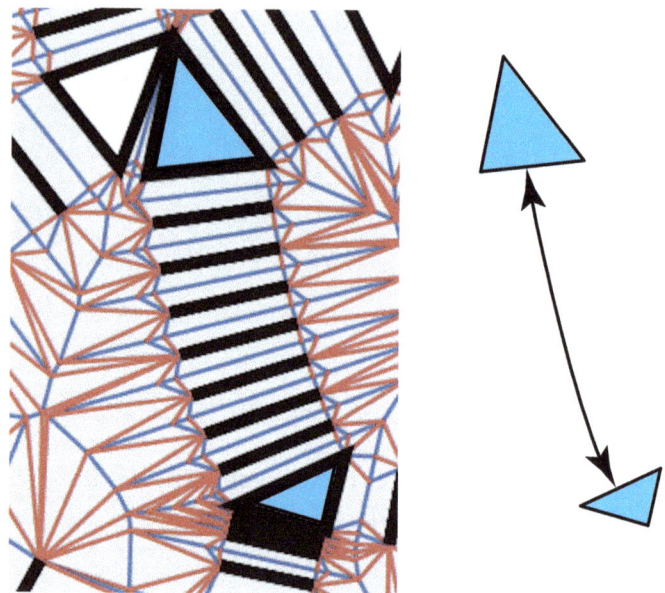

Figure 10.12 Accordion fold to connect two edges. Detail from Figure 10.7.

> **Exercise 10.4 [Practice.] Accordion Pleats**
>
> Revisit the accordion pleat folding of eight 1×1 stamps in Figure 2.2. Using this as a guide, how many internal M/V creases are needed to accordion fold a length L down to height h?

10.3. ORIGAMIZER Algorithm

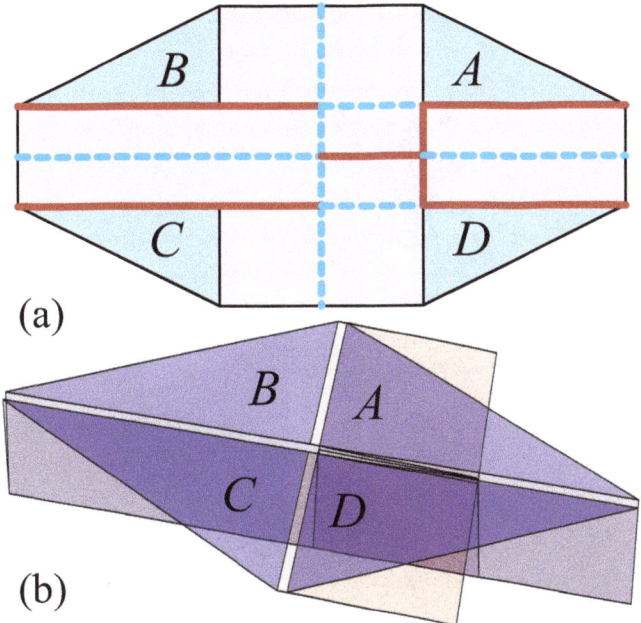

Figure 10.13 Degree-4 vertex-tuck. (a) A,B,C,D are faces of $\mathcal{P}$. (b) 3D folding following the indicated M/V assignments.

Figure 10.14 Detail from Figure 10.7.

Figure 10.15 Metal Bunny following ORIGAMIZER crease pattern. [Reprinted by permission of the photographer, Tomohiro Tachi.]

Vertex-tucking Even with the complication of avoiding paper collisons, edge-tucking is simpler than vertex-tucking. All the faces incident to a vertex v of $\mathcal{P}$ must be drawn together, at the same time as collapsing the separation between the edges incident to v. Consider a simple example: four congruent coplanar right triangles incident to a degree-4 vertex. Figure 10.13(a) shows that creases meeting at right angles suffices, with the M/V assignments shown. The extra paper folds to rectangles on the tuck side: see Figure 10.13(b).

The reality of a noncontrived example is much more complicated. In the cube crease pattern (Figure 10.10), vertex-tucking involves the three squares meeting at a vertex. For example, faces F, L, B meet at v_1 in a nonconvex 10-gon which collapses and folds to tuck underneath. Similarly, faces T, K, R meet at v_7 again in a nonconvex 10-gon.

The bunny crease pattern in Figure 10.14 connects a degree-7 vertex v surrounded by a 14-gon. In this case, the seven edges incident to v are all handled by single bisectors; no accordion pleating is needed. But ORIGAMIZER needs a spider's web of M/V creases to fold the vertex paper to tuck underneath. (Note that Maekawa's Theorem 3.1 is satisfied throughout this tangle: $|M-V| = 2$.) This is one of the more complicated steps of the algorithm, employing a construct called the ***Voronoi diagram***. It would require a long diversion to describe this accurately, so instead we end our description of the algorithm here by reemphasizing the generality of Theorem 10.1: ORIGAMIZER folds a piece of paper to *any* polyhedron.

As an encore, we close this chapter with an image of a (somewhat coarser) Stanford Bunny model of 164 triangles: see Figure 10.15. The model was based on a crease pattern created by ORIGAMIZER, which was then folded from a single 4-ft × 4-ft, thin aluminum sheet, with laser-cut holes to avoid vertex-tucking, judged too complicated to fold in metal. For example, the central convex 14-gon in Figure 10.14 would be removed, and similarly for every vertex.

Both origami design and the Stanford Bunny have come a long way in the past 30 years!

10.4 Technical Notes

Sec. 10.2: Strip Algorithm Presentation based on Demaine and O'Rourke (2007, Sec. 15.1).

Sec. 10.3: ORIGAMIZER Algorithm Algorithm: Tachi (2009). Proof: Demaine and Tachi (2017).

Sec. 10.3.1: Weaknesses Overcome Drawn from Demaine and Tachi (2017).

Sec. 10.3.2: The ORIGAMIZER Algorithm In addition to the two main publications on ORIGAMIZER, lectures by Erik Demaine on ORIGAMIZER are freely available via *MIT OpenCourseWare*.

The metal bunny was built at MIT in 2011 by Ken Cheung, Erik Demaine, Martin Demaine, and Tomohiro Tachi.

Appendix A

Beyond: Topics Not Covered

To fully live up to the title of this book, *The Mathematics of Origami*, would require a much longer book, and one employing more advanced mathematics. Origami touches a wealth of mathematical topics I chose not to cover. Listed below are a dozen such topics, each with a reference or two to seed further exploration.

(1) Origami Axioms, constructible numbers: Demaine and O'Rourke (2007, Ch. 19).

(2) Origami Algebra: Hull (2020, Ch. 3).

(3) Angle trisection: Demaine and O'Rourke (2007, p. 287), Hull (2020, Sec. 1.3). Lang's angle quintisection: Hull (2020, Sec. 4.2).

(4) Combinatorics of flat-foldings: Hull (2020, Ch. 7).

(5) Thick origami: Tachi (2011), Zhu and Filipov (2024).

(6) Origami flasher dynamics: Lang et al. (2016).

(7) Lang's TREEMAKER: Lang (2012, Ch. 11), Demaine and O'Rourke (2007, Ch. 16), Hull (2020, Sec. 8.3).

(8) Twists and Twist Tiles: Lang (2017, Ch. 3,4).

(9) Primal–dual tesselations: Lang (2017, Ch. 6).

(10) Self-folding robots: Felton et al. (2014), Gu et al. (2023).

(11) Margulis napkin folding problem: Lang (2017, p. 144), Hull (2020, p. 172).

(12) Folding a sphere: Lang (2017, Sec. 10.1.2). Wrapping a sphere: Demaine et al. (2007).

Enjoy exploring!

Appendix B

Solutions to Exercises

B.1 Chapter 1 Exercises

There are no Exercises in Chapter 1. In the following sections, each exercise solution is numbered the same as the exercise in which it appeared in the associated chapter.

B.2 Chapter 2 Exercises

> **Exercise 2.1 Solution: Exponential/Polynomial**
>
> Take the logarithm base 2:
> $$\log_2(2^n) = \log_2(n^{100})$$
> $$n = 100 \log_2(n)$$
> $$n/\log_2(n) = 100.$$
> There is no easy way to solve this equation exactly, but one can estimate by assuming $n = 2^k$ so that $\log_2(n) = k$. Suppose $n = 2^{10} = 1024$. Then $n/\log_2(n) = 1024/10 = 102.4$, close to 100. So n is approximately 1000. Numerical calculation shows that $n \approx 996$.

> **Exercise 2.2 Solution: $n!$ Growth**
>
> First fix $n = 4$. In the expression $4! = 4 \cdot 3 \cdot 2 \cdot 1$, replace all numbers by 2, so forming
> $$4! = 24 > 2 \cdot 2 \cdot 2 \cdot 2 = 2^4 = 16\ .$$
> For larger values of n, we continue to replace numbers larger than 2 by 2, so 2^n is a lower bound. For the upper bound, replace each number in $n!$ with n, and so forming n^n:
> $$4! = 24 < 4 \cdot 4 \cdot 4 \cdot 4 = 4^4 = 256\ .$$

169

Exercise 2.3 Solution: $n!$

Suppose $abc\cdots q$ is a particular ordering of $n-1$ objects, and x is the nth object. Then place x before a, then between a and b, etc.:

$$(\mathbf{x}abc\cdots q),\, (a\mathbf{x}bc\cdots q)\, \ldots\, (abc\cdots q\mathbf{x}).$$

This converts each ordering of $n-1$ objects into n different orderings of n objects. This is why $n! = n\cdot(n-1)!$.

Exercise 2.4 Solution: 4×2 Map

Building off the 3×2 map folding in Figure 2.13, follow these steps: V-fold squares 7/8 on top of 5/6. V-fold this on top of 3/4. V-fold in half along horizontal. M-fold in half again. This leads to the bottom-to-top ordering (1,2,4,8,6,5,7,3).

B.3 Chapter 3 Exercises

Exercise 3.1 Solution: Not Flat Foldable

Looking ahead, these are the obstructions to flat-foldability:

(a) The four mountain creases violate Maekawa's Theorem 3.1.

(b) The degree-3 vertex violates the Even-Degree Lemma 3.1.

(c) The degree-4 vertex violates the Local-Min Lemma 3.2.

Exercise 3.2 Solution: Maekawa's Theorem

Two solutions are shown in Figure B.1, both satisfying $M = 4$, $V = 2$.

Exercise 3.3 Solution: Local-Min Ties

(a) YES, the pattern can fold flat.

(b) It does not violate the Local-Min Lemma 3.2, which requires a strict local min. Here $45° = 45° < 90°$. Even with a tie to just one side, the collision of paper can be avoided.

B.4. Chapter 4 Exercises

Exercise 3.4 Solution: Local-Min and Kawasaki's Theorem

(a) Kawasaki's Theorem 3.2 is satisfied:
$$70° + 70° + 40° = 40° + 70° + 70° = 180°.$$

(b) The Local-Min-Lemma 3.2 constrains the M/V assignment. In fact the crease pattern in Figure 3.6(a) *is* flat-foldable, but not with the M/V assignment illustrated.

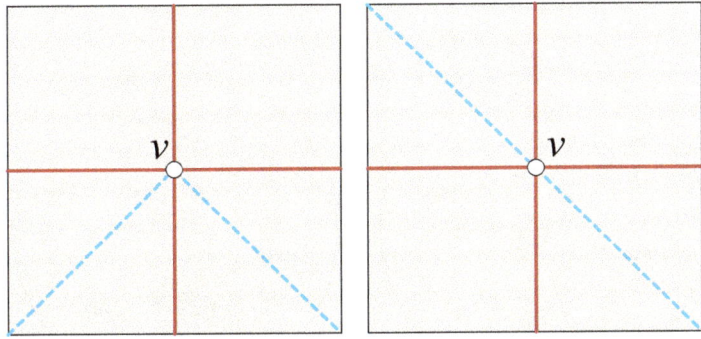

Figure B.1 Adding two V creases satisfies Maekawa's Theorem and permits these pattern to fold flat.

Exercise 3.5 Solution: Exponentially Many M/V

See Figure B.2. Note that the 45° angle is a local min—90° > 45° < 90°—and so by the Local-Min Lemma 3.2, it must be delimited by one M and one V crease.

B.4 Chapter 4 Exercises

Exercise 4.1 Solution: Set Partition

Sets (a) and (b) are nearly immediate. For (c), first note the sum of all the numbers in S is 28, then look for combinations that sum to 14.

(a) $1 + 2 = 3$.

(b) $1 + 4 = 2 + 3$.

(c) $3 + 5 + 6 = 1 + 2 + 4 + 7$.

Exercise 4.2 Solution: 3-SAT

Either $x, y, z = F, T, F$ or $x, y, z = F, F, T$.

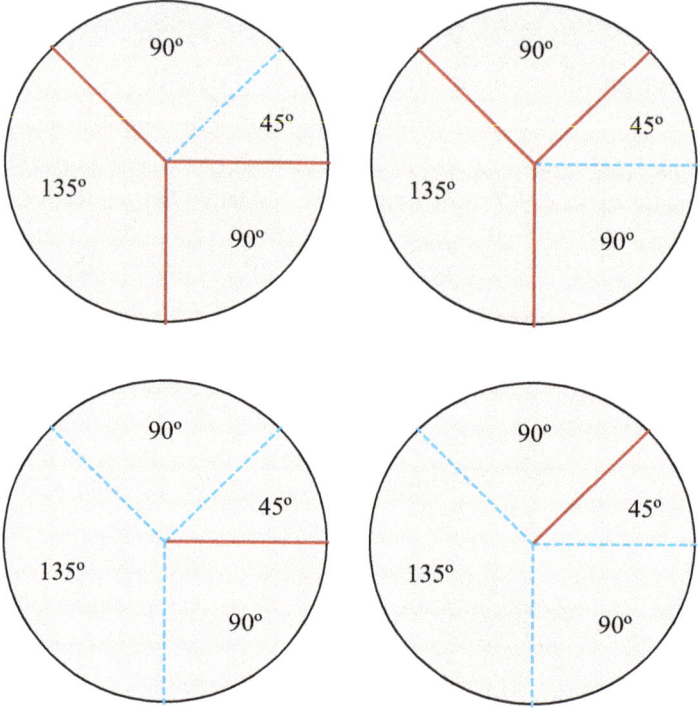

Figure B.2 Four M/V patterns that each fold flat.

> ### Exercise 4.3 Solution: Splitter Self-Crossing
>
> Use three separate folds performed in sequence:
>
> (1) M-fold along yz, 180° (flat) to 0° (fully creased).
>
> (2) V-fold along xy, 180° to 0°.
>
> (3) V-fold along xz, 180° to 0°.

> ### Exercise 4.4 Solution: Splitter with Surround
>
> A trapezoid of exterior paper folds above y, and another trapezoid below z. A triangle of exterior paper folds to the right of x.

> ### Exercise 4.5 Solution: Rule 110
>
> See Figure B.3 for five generations of evolution starting from 101.

B.4. Chapter 4 Exercises

Figure B.3 Five generations evolving from 1 0 1.

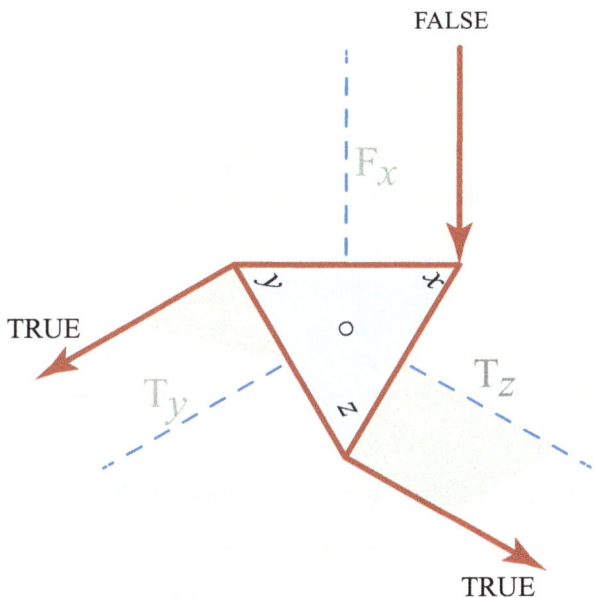

Figure B.4 Figure 4.14(a) when the input is FALSE.

Exercise 4.6 Solution: Triangle Twist

The triangle twists 60° clockwise. See Figure B.4.

B.5 Chapter 5 Exercises

Exercise 5.1 Solution: Cube from 3×3 Square

(a) In the 3×3 square pattern, each red diagonal has length $\sqrt{2}$. So when the box sides are rotated close to vertical, the tip p of a fin projects to $p_\perp$ on the base of the cube beyond the base center x. See Figure B.5. So the four fins would collide and penetrate one another.

(b) The collisions can be avoided by first folding one corner square (say, rightmost, topmost) and rotating its doubled triangle fin to the back face. And only then fold the next corner square (leftmost, topmost), and so on. So there is never more than one fin in the center at a time: Each is rotated away before the next is folded.

Exercise 5.2 Solution: Miura Area

Completely opened:
$$mn\, 2\sqrt{3},$$
because each Miuri-unit consists of four rhombs and so eight equilateral triangles, each of area $\sqrt{3}/4$.

Completely folded:
$$(2m+1)\sqrt{3}/4,$$
because it is composed of $m+1$ right-side-up equilateral triangles, and m upside-down. Note the collapsed height is $\sqrt{3}/2$ independent of n.

Exercise 5.3 Solution: Plus-Sign Proof

(1) If $\alpha = 180°$, then $\gamma = \alpha = 180°$ and $\beta = \delta$.

(b) If $\alpha = 0°$, then $\gamma = \alpha = 0°$ and $\beta = -\delta$, one edge is M and the other V.

Exercise 5.4 Solution: MM Configuration

The upper circle intersection y is the reflection of x in the xy-plane, corresponding to the c position when ao and oc are two collinear M-folds, violating the stipulation that oc is a V-fold.

B.5. Chapter 5 Exercises

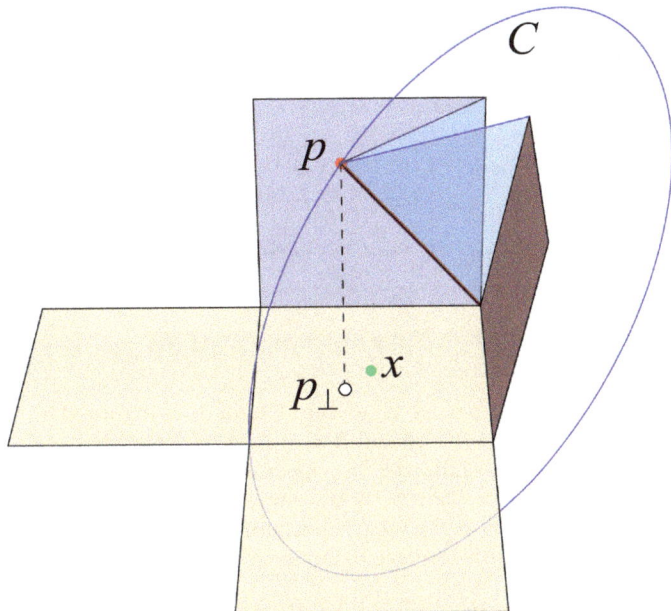

Figure B.5 Point p travels on circle C of radius $\sqrt{2}$ centered on a cube corner. $p_\perp$ is the projection of p onto the cube base, whose center is x.

Exercise 5.5 Solution: $\theta \to 90°$

As $\theta \to 90°$, the rhombs become squares, and the edges of the pattern lie along a unit grid. The Miura units each become plus signs. The major folds produce a rectangle strip, and the minor folds "accordion fold" the strip into a zig-zag. So the whole pattern folds into a 1×1 square!

Exercise 5.6 Solution: Flat Folding Degree-4 Vertex

The two crease patterns in Figure 5.21(a) and (b) are identical, so Kawasaki's Theorem 3.2, which relies solely on the sector angles incident to a vertex, can be verified by the face angle sums:

$$90° + 90° = 45° + 135° = 180° \ .$$

Makawa's Theorem 3.1 is easily checked: 3M + 1V or 1M + 3V. There is a local min in both the (a) or (b) patterns—$90°, 45°, 90°$—but the $45°$ angle is delimited by M to one side and V to the other. So the Local-Min Lemma 3.2 is satisfied.

Exercise 5.7 Solution: Square Twist Area

Let L_{open} be the side length of the square paper when fully opened, and L_{fold} when fully folded flat:

$$L_{\text{open}} = (2 + \sqrt{2}) \approx 3.41$$
$$L_{\text{fold}} = (2 + \sqrt{2}/2)/2 + (1 - \sqrt{2}/2) = 2$$
$$L_{\text{fold}}^2 / L_{\text{open}}^2 = 6 - 4\sqrt{2} \approx 0.343.$$

Exercise 5.8 Solution: Nonrigid Pattern

Focus on the highlighted vertex v_1 of the diamond in Figure B.6. We know from Theorem 5.2 that $\alpha_1 = -\gamma_1$ and $\beta = \delta$. And we know that $\alpha_1 > \beta_1$ from Equation (5.2). The arrows indicate ">," meaning "flatter than" because these are dihedral angles. Notice that β_1 is the same as α_4, so $\alpha_1 > \alpha_4$. Similarly, from vertex v_2 we have $\alpha_2 > \beta_2 = \alpha_1$. Continuing counterclockwise around to v_3 and v_4 leads to the central square being assigned four α dihedrals satisfying

$$\alpha_4 > \alpha_3 > \alpha_2 > \alpha_1 > \alpha_4 .$$

This cyclic "race condition" shows rigid folding is impossible.

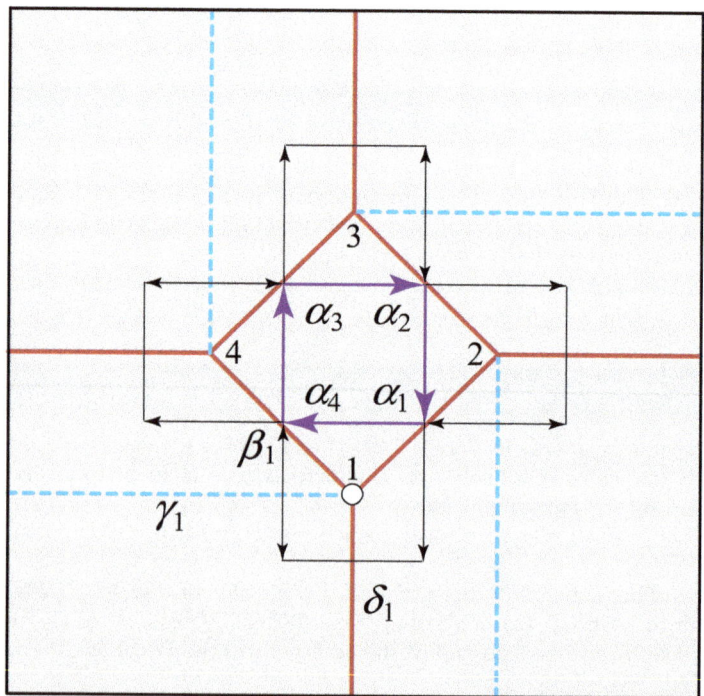

Figure B.6 Dihedral constraints around the four vertices of crease pattern M^4.

B.6 Chapter 6 Exercises

Exercise 6.1 Solution: Collapsing to a Line

Every tree can collapse to a line because a tree has no cycles. For example, a triangle constitutes a cycle of three edges, and clearly a triangle cannot be flattened without changing its edge lengths.

Exercise 6.2 Solution: Waterbomb Cube

As can be seen by the central subsquare in the crease pattern in Figure 6.5(b), each face of the cube is $\frac{1}{4} \times \frac{1}{4}$.

Exercise 6.3 Solution: Bisectors Meet at Incenter

This is Proposition 4, Book IV in Euclid. The angle bisector from a vertex a is perpendicularly equidistant from the two edges incident to that vertex. The point x where a and b's angle bisectors meet is perpendicularly equidistant from the edges incident to the third vertex c, for those two edges are also incident to a and b. So the bisector of c hits x.

Exercise 6.4 Solution: Rabbit Ear Fold Flat

Let α, β, γ be the triangle angles at a, b, c. Kawasaki's Theorem 3.2 is satisfied when $\theta_1 + \theta_3 = \theta_2 + \theta_4 = 180°$:

$$\theta_1 = 180° - \alpha/2 - 90°$$
$$\theta_2 = 180° - \beta/2 - 90°$$
$$\theta_3 = 180° - \beta/2 - \gamma/2$$
$$\theta_4 = 180° - \gamma/2 - \alpha/2$$

$$\theta_1 + \theta_3 = 270° - \frac{1}{2}(\alpha + \beta + \gamma) = 270° - 90° = 180°$$
$$\theta_2 + \theta_4 = 270° - \frac{1}{2}(\alpha + \beta + \gamma) = 270° - 90° = 180°.$$

So Kawasaki's Theorem holds.

Exercise 6.5 Solution: Two Tangents Theorem

Let axb' be a right triangle with hypotenuse ax. The segments xb' and xd' have the same length. Then the hypotenuse-leg congruence theorem shows the two triangles axb' and axd' are congruent.

Exercise 6.6 Solution: Quadrilateral → Triangle

Let $\alpha, \beta, \gamma, \delta$ be the four angles of the quadrilateral in counterclockwise order. The sum $\alpha + \beta + \gamma + \delta = 360°$ (because a diagonal splits the quadrilateral into two triangles). If $\alpha + \beta < 180°$, then the extension of the two edges before α and after β meet and form a triangle, satisfying the claim.

Suppose instead that $\alpha + \beta \geq 180°$. Then it must be that $\gamma + \delta \leq 180°$. If $\gamma + \delta < 180°$, then again the extensions of two opposite edges meet. If $\gamma + \delta = 180°$, then both $\alpha + \beta$ and $\gamma + \delta$ sum to exactly $180°$ and the quadrilateral is a rectangle, excluded from the claim.

Exercise 6.7 Solution: Nontangential

One can see visually, or by rough measurement, that $|ab| + |cd|$ is about 20 percent longer than $|bc| + |da|$. So by Pitot's Theorem 6.1, the quadrilateral cannot be tangential.

Exercise 6.8 Solution: Straight Skeleton

(a) Six to nine straight-skeleton edges. The straight skeleton of a regular hexagon has just six edges (see ahead to Figure 7.4), but if all internal nodes are degree-3, such an irregular hexagon has nine edges.

(b) The general formula is anywhere from n (for a regular n-gon) to $2n - 3$ skeleton edges.

B.7 Chapter 7 Exercises

Exercise 7.1 Solution: 1-Cut Square

Refer to Figure B.7. M-fold diagonal bd, then bring b to d by folding the M and V creases along the ac diagonal. Now the four boundary edges of the square lie on top of one another, with the square interior above and the exterior paper below, ready for 1-cut.

Exercise 7.2 Solution: Folded Layers

The maximum thickness is six layers, for example, at a point slightly below x.

B.7. *Chapter 7 Exercises* 179

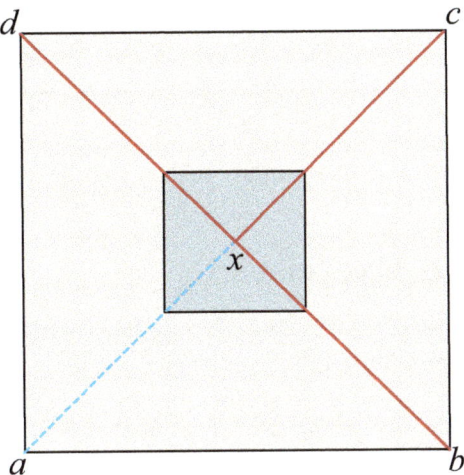

Figure B.7 Fold & 1-Cut a square.

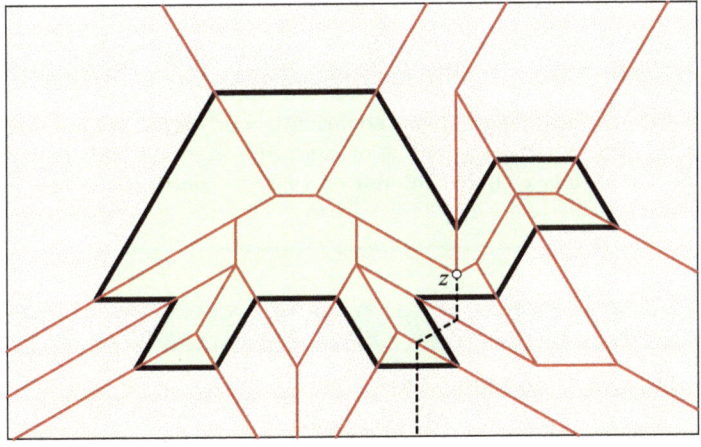

Figure B.8 Wandering perpendicular originating at z.

> **Exercise 7.3 Solution: Wandering Perpendiculars**
>
> See Figure B.8: The only choice at vertex z is downward. There are other vertices in the figure that also force wandering perpendiculars.

> **Exercise 7.4 Solution: Covering by Disks**
>
> A rectangle of height 1 and width w needs $2(w+1)$ disks. In Figure B.9, $w = 10$ and 22 disks are needed. So choose $2(w + 1) > m$, or $w > m/2 + 1$.

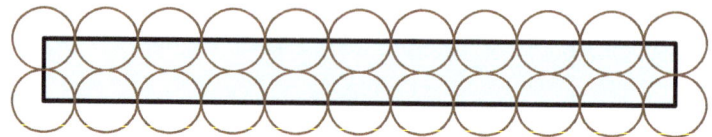

Figure B.9 A thin rectangle needs many disks to cover its edges.

B.8 Chapter 8 Exercises

Exercise 8.1 Solution: Area of Lune

The lune is composed of two circle **sectors**, one for each circle. Using formulas for calculating the specific sector areas illustrated in Figure 8.4 leads to lune area

$$\frac{2\pi}{3} - \frac{\sqrt{3}}{2} \approx 1.23.$$

Exercise 8.2 Solution: Curvature of Parabola

A parabola with focal length f has equation $y = x^2/(4f)$, so the focal length of $y = x^2/4$ is $f = 1$. So the radius of curvature is $r = 2$ and the curvature is $\kappa = \frac{1}{2}$. The radius-2 circle centered on $(0,2)$ osculates the parabola at its vertex $(0,0)$.

Exercise 8.3 Solution: Circle Parametric Equations

For t in the range $0°$ to $360°$:

$$x(t) = \cos(t)$$
$$y(t) = \sin(t).$$

Check: $x^2 + y^2 = \cos^2 + \sin^2 = 1$.

Exercise 8.4 Solution: Parabola Parametric Equation

For t in the range 0 to ∞:

$$x(t) = 2t$$
$$y(t) = t^2.$$

Check: $y = x^2/4 = (2t)^2/4 = t^2$.

Exercise 8.5 Solution: Circle Tangent Bisection

The focus of a circle is its center. Arrange so that t_1 and t_2 lie on the same vertical line. Then each is the reflection of the other with respect to the horizontal line through f and v. So the triangles fvt_1 and fvt_2 are similar, and fv bisects the angle $\angle t_1 f t_2$.

B.9 Chapter 9 Exercises

Exercise 9.1 Solution: Waterbomb Asymmetric

Refer to Figure B.10. Three large right triangles meet at the origin o, where the V-folds form an xyz coordinate system. Two back-to-back copies of the small right triangle oaz jut into the interior.

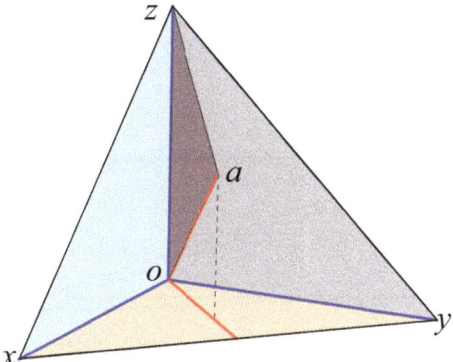

Figure B.10 Dihedrals along red segments are 0° and 180°.

Exercise 9.2 Solution: Cut Trees

See Figure B.11. The cut tree for net **B** is easier to see because it is a path—no branching.

Exercise 9.3 Solution: Convex Hull Perimeter

In Figure 9.5:
- (a) $4 + 2\sqrt{2} + 2\sqrt{5} \approx 11.3$.
- (b) $5 + 3\sqrt{2} + \sqrt{5} \approx 11.5$.

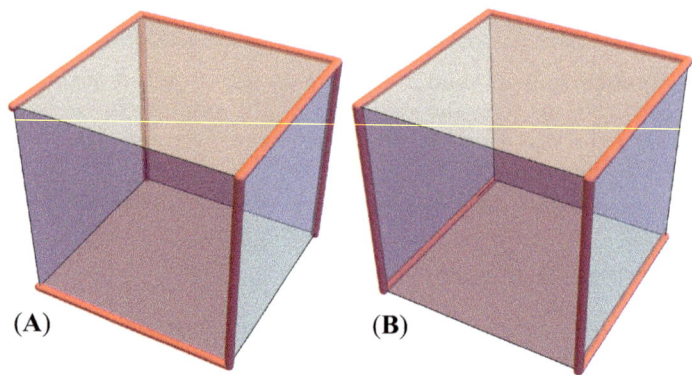

Figure B.11 Cut trees for the nets **A** and **B** in Figure 9.4.

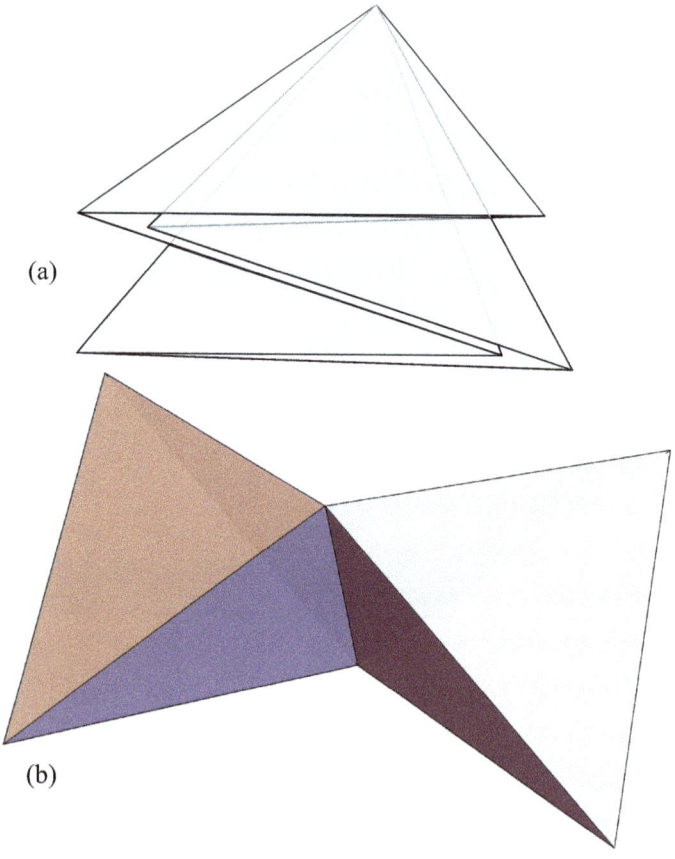

Figure B.12 (a) Hexagon folded flat. (b) Hexagon rigidly folded MMVMMV.

B.10. Chapter 10 Exercises

Exercise 9.4 Solution: Pinched Hexagon

(a) YES, six triangles can fold to a zig-zag stack of equilateral triangles. The hexagon boundary forms a Z-shape as depicted in Figure B.12(a).

(b) See Figure B.12(b). In the limit the two V-creases touch forming a bow-tie shape.

Exercise 9.5 Solution: Plus-Sign Congfiguration Space

See Figure B.13. There are two pathways: α varies along the horizontal axis while $\beta = 0$, and β varies along the vertical axis while $\alpha = 0$.

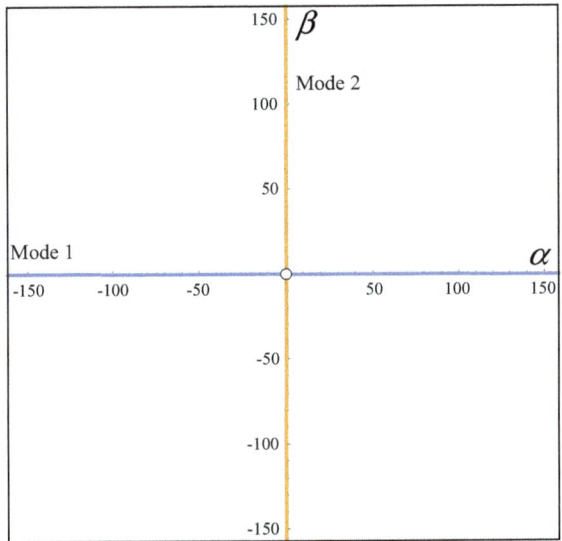

Figure B.13 Two pathways in the plus-sign configuration space.

B.10 Chapter 10 Exercises

Exercise 10.1 Solution: Latin Cross

See Figure B.14 for one (of several) solutions.

Exercise 10.2 Solution: Dodecahedron Hamiltonian Cycle

See Figure B.15.

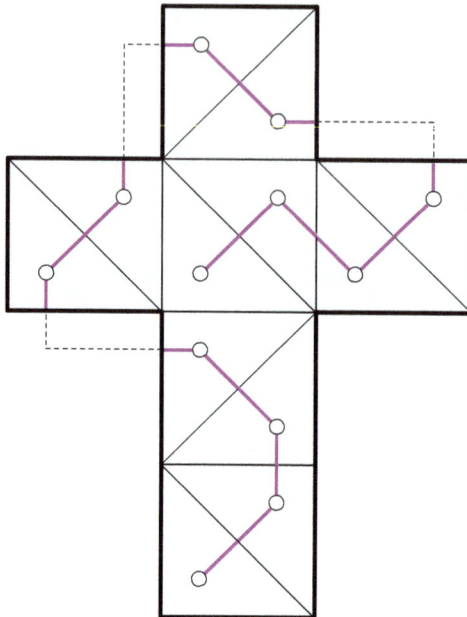

Figure B.14 Hamiltonian triangulation of the Latin cross net.

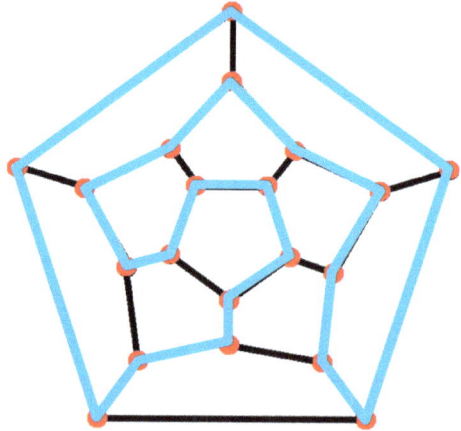

Figure B.15 A Hamiltonian cycle (blue) on the edges of the dodecahedron.

Exercise 10.3 Solution: Non-Hamiltonian Refinement

See Figure B.16.

B.10. Chapter 10 Exercises

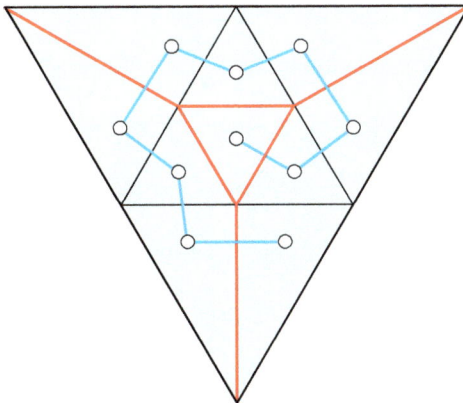

Figure B.16 A Hamiltonian path after splitting triangles in Figure 10.3 with chords (red).

Exercise 10.4 Solution: Accordion Pleats

It's easiest to assume that h divides L. Then $L/h - 1$ creases suffice to reduce L to height h. For example, to achieve $h = 2$ for $L = 14$, $14/2 - 1 = 6$ creases are needed.

References

Abel, Z., E. D. Demaine, M. L. Demaine, J.-i. Itoh, A. Lubiw, C. Nara, and J. O'Rourke (2014). Continuously flattening polyhedra using straight skeletons. In *Proc. 30th Ann. Symp. Comput. Geom.*, pp. 396–405. Association for Computing Machinery.

Akitaya, H. A., K. C. Cheung, E. D. Demaine, T. Horiyama, T. C. Hull, J. S. Ku, T. Tachi, and R. Uehara (2016). Box pleating is hard. In *Discrete Comput. Geom. Graphs: 18th Japan Conf. JCDCGG 2015, Kyoto, Japan.*, pp. 167–179. Springer.

Akitaya, H. A., E. D. Demaine, T. Horiyama, T. C. Hull, J. S. Ku, and T. Tachi (2020). Rigid foldability is NP-hard. *J. Comput. Geom. 11*(1), 93–124.

Azam, A., T. G. Leong, A. M. Zarafshar, and D. H. Gracias (2009). Compactness determines the success of cube and octahedron self-assembly. *PloS one 4*(2), e4451.

Bern, M., E. D. Demaine, D. Eppstein, and B. Hayes (2002). A disk-packing algorithm for an origami magic trick. In *Origami^3: Proc. 3rd Internat. Meet. Origami Sci., Math., Educ. (2001)*, pp. 17–28. A K Peters.

Bern, M. and B. Hayes (1996). The complexity of flat origami. In *Proc. 7th ACM-SIAM Sympos. Discrete Algorithms*, pp. 175–183. Association for Computing Machinery.

Cepelewicz, J. (2024). How to build an origami computer. *Quanta Magazine*. 30 January 2024.

Cook, M. (2004). Universality in elementary cellular automata. *Complex Systems 15*(1), 1–40.

Cormen, T. H., C. E. Leiserson, R. L. Rivest, and C. Stein (2022). *Introduction to Algorithms* (4th ed.). MIT Press.

Demaine, E. D., M. L. Demaine, V. Hart, G. N. Price, and T. Tachi (2011a). (Non) existence of pleated folds: How paper folds between creases. *Graphs Combinatorics 27*, 377–397.

Demaine, E. D., M. L. Demaine, J. Iacono, and S. Langerman (2007). Wrapping the Mozartkugel. In *24th Euro. Workshop Comput. Geom. (EuroCG 2007), Graz, Austria*, pp. 14–17.

Demaine, E. D., M. L. Demaine, and D. Koschitz (2011b). Reconstructing David Huffmans legacy in curved-crease folding. *Origami^5: Proc. 5th Internat. Meet. Origami Sci., Math., Educ. (2011)*, pp. 39–52. CRC Press.

Demaine, E. D., M. L. Demaine, D. Koschitz, and T. Tachi (2015). A review on curved creases in art, design and mathematics. *Symmetry: Culture Sci. 26*(2), 145–161.

Demaine, E. D., K. Mundilova, and T. Tachi (2022). Locally flat and rigidly foldable discretizations of conic crease patterns with reflecting rule lines. In *Internat. Conf. Geom. Graphics*, pp. 185–196. Springer.

Demaine, E. D. and J. O'Rourke (2007). *Geometric Folding Algorithms: Linkages, Origami, Polyhedra*. Cambridge University Press.

Demaine, E. D. and T. Tachi (2017). Origamizer: A practical algorithm for folding any polyhedron. In *33rd Internat. Symp. Comput. Geom.* Schloss Dagstuhl-Leibniz-Zentrum für Informatik.

Devadoss, S. and J. O'Rourke (2025). *Discrete and Computational Geometry* (2 ed.). Princeton University Press.

Eppstein, D. (2024). Computational complexities of folding. arXiv:2410.07666.

Farnham, J., T. C. Hull, and A. Rumbolt (2022). Rigid folding equations of degree-6 origami vertices. *Proc. Royal Soc. A 478*(2260), 20220051.

Felton, S., M. Tolley, E. D. Demaine, D. Rus, and R. Wood (2014). A method for building self-folding machines. *Science 345*(6197), 644–646.

Fuchs, D. and S. Tabachnikov (1999). More on paperfolding. *Amer. Math. Monthly 106*(1), 27–35.

Gu, H., M. Möckli, C. Ehmke, M. Kim, M. Wieland, S. Moser, C. Bechinger, Q. Boehler, and B. J. Nelson (2023). Self-folding soft-robotic chains with reconfigurable shapes and functionalities. *Nat. Commun. 14*(1), 1263.

Horiyama, T. and W. Shoji (2011). Edge unfoldings of Platonic solids never overlap. In *Proc. 23rd Canad. Conf. Comput. Geom.*, 65–70.

Houdini, H. (1922). *Paper Magic*, pp. 176–177. E. P. Dutton & Company. Reprinted by Magico Magazine.

Huffman, D. A. (1976). Curvature and creases: A primer on paper. *IEEE Trans. Comput. C-25*, 1010–1019.

Hull, T. C. (2020). *Origametry: Mathematical Methods in Paper Folding*. Cambridge University Press.

Hull, T. C. and M. T. Urbanski (2018). Rigid foldability of the augmented square twist. arXiv:1809.04899 [math.MG].

Hull, T. C. and I. Zakharevich (2023). Flat origami is Turing complete. arXiv:2309.07932.

Itoh, J.-i., C. Nara, and C. Vîlcu (2012). Continuous flattening of convex polyhedra. In *Comput. Geom.*, Volume 7579 of Lecture Notes in Computer Science, pp. 85–97. Springer.

Jungck, J. R., S. Brittain, D. Plante, and J. Flynn (2022). Self-assembly, self-folding, and origami: Comparative design principles. *Biomimetics 8*(1), 12.

Kuribayashi, K., K. Tsuchiya, Z. You, D. Tomus, M. Umemoto, T. Ito, and M. Sasaki (2006). Self-deployable origami stent grafts as a biomedical application of Ni-rich TiNi shape memory alloy foil. *Mater. Sci. Eng: A 419*(1-2), 131–137.

Lang, R. J. (1996). A computational algorithm for origami design. In *Proc. 12th Ann. Sympos. Comput. Geom.*, Philadelphia, PA, pp. 98–105.

Lang, R. J. (2003). *Origami Design Secrets: Mathematical Methods for an Ancient Art* (1st ed.). A K Peters.

Lang, R. J. (2012). *Origami Design Secrets: Mathematical Methods for an Ancient Art* (2nd ed.). A K Peters/CRC Press.

Lang, R. J. (2017). *Twists, Tilings, and Tessellations: Mathematical Methods for Geometric Origami*. A K Peters/CRC Press.

Lang, R. J., S. Magleby, and L. Howell (2016). Single degree-of-freedom rigidly foldable cut origami flashers. *J. Mech. Robot. 8*(3), 031005.

Lukasheva, E. (2021). *Curved Origami: Unlocking the Secrets of Curved Folding in Easy Steps*. New Origami Publishing.

Ma, J., H. Feng, Y. Chen, D. Hou, and Z. You (2020). Folding of tubular waterbomb. *Research* Vol. 2020, Article ID: 1735081.

Morgan, T. D. (2012). *Map folding*. M. Eng. thesis, Massachusetts Institute of Technology.

Mundilova, K. (2019). On mathematical folding of curved crease origami: Sliding developables and parametrizations of folds into cylinders and cones. *Comput.-Aided Des. 115*, 34–41.

Mundilova, K. (2024). *Gluing and Creasing Paper along Curves: Computational Methods for Analysis and Design*. Ph. D. thesis, Massachusetts Institute of Technology.

References

Mundilova, K. and T. Wills (2018). Folding the Vesica Piscis. In *Proc. Bridges 2018: Math., Art, Music, Arch., Educ., Cult.*, pp. 535–538.

O'Rourke, J. (2011). *How to Fold It: The Mathematics of Linkages, Origami, and Polyhedra*. Cambridge University Press.

O'Rourke, J. (2022). *Pop-Up Geometry: The Mathematics Behind Pop-Up Cards*. Cambridge University Press.

O'Rourke, J. (2024). Review of Origametry: Mathematical Methods in Paper Folding, by Thomas Hull. *Math. Intell.* 47(1).

Randall, C. L., E. Gultepe, and D. H. Gracias (2012). Self-folding devices and materials for biomedical applications. *Trends Biotechnol.* 30(3), 138–146.

Resch, R. D. (1974). The space curve as a folded edge. In *Comput. Aided Geom. Des.*, pp. 255–258. Elsevier.

Santangelo, C. D. (2017). Extreme mechanics: Self-folding origami. *Ann. Rev. Conden. Ma. P.* 8(1), 165–183.

Sipser, M. (2012). *Introduction to the Theory of Computation* (3rd ed.). Cengage Learning.

Stern, M., M. B. Pinson, and A. Murugan (2017). The complexity of folding self-folding origami. *Phys. Rev. X* 7(4), 041070.

Tabachnikov, S. (2014). Dragon curves revisited. *Math. Intell.* 36, 13–17.

Tachi, T. (2009). Origamizing polyhedral surfaces. *IEEE Trans. Vis. Comp. Graphics* 16(2), 298–311.

Tachi, T. (2011). Rigid-foldable thick origami. *Origami^5: Proc. 5th Internat. Meet. Origami Sci., Math., Educ. (2011)*, pp. 253–264. CRC Press.

Tachi, T. (2013). Composite rigid-foldable curved origami structure. *Proc. Transformables*, 18–20.

Tachi, T. and T. C. Hull (2017). Self-foldability of rigid origami. *J. Mech. Robot.* 9(2), 021008.

Uehara, R. (2020). *Introduction to Computational Origami*. Springer.

Wolfram, S. (2002). *A New Kind of Science*. Wolfram Media Inc.

Zhu, Y. and E. T. Filipov (2024). Large-scale modular and uniformly thick origami-inspired adaptable and load-carrying structures. *Nat. Commun.* 15(1), 2353.

Index

2-colored, 24
3-SAT, 37–38
   NAE 3-SAT, 40
   solvers, 38
   truth assignment, 37
3-Sum, 36
3-space, 1, 3
4D rotation, 120

Abrashi, Arijan, 101, 104
alternating sum, 27–31, 138
   partial, 31
∧, AND, 37, 51
angle
   ∠, 137
   dihedral, 3, 4
   fold, 4
   supplementary, 3, 4
   trisection, 168
angle bisector, 108
   in quadrilateral, 95
   straight skeleton, 108
   in triangle, 92, 177
   of two edges, 108, 113

base, origami
   fish, 87
   traditional, 87
   uniaxial, 88–92, 95, 100, 118
   uniaxial properties, 90–92, 102
   waterbomb, 89, 91, 142
   windmill, 87
box-pleat pattern, 40, 46, 58
boxes, mathematics, x, 1

calculus, ix, 129, 130
cellular automaton

1D, 50, 52
2D, 48, 50
Chapman, Paul, 49
characterization, xi, 27, 31–33, 98
circle packing, 87, 94, 107
circle/river method, 92, 99–102
combinatorics, 6, 10
complexity
   computational, ix, 18, 33, 36
   hierarchy, 36, 37
   zoo, 36
computational origami, ix, 33, 153
cone, *vesica piscis*, 124, 125
configuration space, 147–151
   distractor paths, 150, 151
convex hull, 146, 181
Conway, John Horton, 48
Cook, Matthew, 51
crease, 1
   conic, 134–136
   curved, 131–133
   major/minor, 67, 77
   optional, 34, 51, 52, 55, 58
   plane, 120
   rolling, 119
crease pattern, 2, 19
cube
   nets, 144, 145
   Origamizer, 161
   waterbomb, 90, 158
curvature, 128
   2D, 128, 129
   3D, 130
   parabola, 129, 180
curve
   Any Curve 2D → 3D Theorem, 133
   Any Curve 3D → 2D Theorem, 132

Index

dragon, 8–10
  Osculating Plane Bisection Theorem, 131
  smooth, 2D, 128, 132, 133
  smooth, 3D, 128, 130
  space-filling, 10
curved-crease origami, 121–140
cylinder
  hexagonal, 63
  tube, 142
  vesica piscis, 124, 132

Degree-4 Folding Theorem, 5, 32, 76, 81, 82, 84, 86, 136, 139, 147, 176
degrees-of-freedom (DOF), 64, 147
  1-DOF, 64–79, 82, 133, 134, 139, 147, 148
  k-DOF, 147
Demaine, Erik D., ix, xii, 66, 107, 121, 123, 127, 159–161, 167
Demaine, Martin L., 107, 121, 123, 127, 167
dihedral angle, 3, 4
  extreme, 69
disk packing, 107, 117
  gaps, 117
disks, tangential, 117
dodecahedron, 4, 144–146, 155, 156, 184
  nets, 146
dragon curve, 6, 8–10

elliptic integral, 125
Euclid, 122, 177
Even-Degree Lemma, 21, 24–27, 31, 33, 56, 170
exercises, x, 169
exponential, 16, 31, 32, 36, 151
  growth, xi, 6, 7, 14, 17, 32, 36
  subexponential, 38
  time, 17, 37, 38, 86, 102

face, 59
  straight skeleton, 113
facet, 59
factorial, 11
flasher, 63, 65, 147, 148, 168
flat folding, 19, 21, 33, 170
  is NP-hard, 38–46
  is Turing-complete, 51–56
  rigid, 85, 86

flattening polyhedra, 119
fold
  accordion, 7, 29, 31, 32, 162–164, 175
  flat vertex, 19–32
  M- (red), 2
  mountain, 2
  pleat, 7, 12, 29, 32, 41, 162–164
  pleat, circular, 127
  pleat, squares, 126
  V- (blue, dashed), 2
  valley, 2
Fold & 1-Cut
  3D polyhedron, 120
  for convex polygon, 109
Fold & 1-Cut Theorem, 105, 118
  disk packing proof, 117–118
  straight skeleton proof, 108–116
fold angle, 4
fractal, 6

gadget, origami, 5, 33, 42, 92
  clause, 42, 46
  cross-over, 42, 45, 53
  eater, 53, 56
  logic-gate, 51
  negation, 42
  splitter, 42, 43, 45, 53
  triangle, 33, 39, 46
  triangle twist, 53–55
  turn, 156
Gosper glider, 49, 51
Gosper, Ralph William Jr., 49
Goucher, Adam P., 49
graph, 89
  dual, 155, 156
Greene, David, 49

half-tangent equation, 71, 75, 76, 79, 80, 84, 150
  multiplier, 76, 79, 80
Hamilton, William Rowan, 155
Hamiltonian cycle, 155, 184
Hamiltonian path, 155, 156
Hamiltonian triangulation, 155
HanaFlex solar array, 63, 64
Houdini, Harry, 105, 120
Huffman, David, 121, 122, 131, 134, 135, 139
Hull, Thomas C., ix, xii, 57, 58
hyperbolic paraboloid, 121, 125–128

icosahedron, 155
if and only if, 26, 40, 46, 99
Ignaut, Brian, 61
∩, intersect, 3

Japanese Space Flyer Unit, 63, 65
jumping frog, 91
Justin, Jacques, 21, 27

Kawasaki's Theorem, xi, 5, 21, 25–33, 38, 39, 43, 76–78, 93, 138, 175, 177
kissing points, 118

Lang, Robert J., ix, xii, 63, 81, 86–88, 104
LIFE, 48–49
local min, 40, 175
  strict, 25, 170
Local-Min Lemma, 21, 25–27, 31, 33, 39–41, 44, 52, 77, 170, 171, 175
log plot, 8, 14
logarithm base, 8, 169
logic gates, *see* gadget, origami
  AND, OR, NOT, 49
  NAND, NOR, 51
Lubiw, Anna, 107
Lukasheva, Ekaterina, 121, 123
lune, 122, 180

M/V assignment, 2, 16, 18, 19
M^2V^2 pattern, 77
M^4 pattern, 77, 83
Maekawa's Theorem, 21–25, 27, 31, 55, 56, 61, 78, 84, 109, 110, 165, 170, 171
map folding, 6, 14, 16, 17, 170
  $n \times 2$, 17, 18
Margulis napkin folding problem, 168
MATHEMATICA, 50
medial axis, 103, 104
Miura map fold, ix, 59, 64, 67, 68, 71, 139
Miura, Koryo, 63, 64
Miura-unit, 66
  Dynamics Theorem, 71
  tesselation, 76, 77, 147, 151
modular origami, 1, 123, 127
molecule, 92–100, 103
  arrowhead, 100

four-circle quadrilateral, 95, 118
gusset, 100
rabbit ear, 92, 94, 108, 118, 148, 177
sawhorse, 99
waterbomb, 94, 142
Mundilova, Klara, xii, 124, 125, 135, 139

n-D rotation, 120
necessary and sufficient, 21, 26, 27, 38
necessary condition, 21, 22, 25, 28, 31, 38
neighborhood, 22, 25
net, 144–146
  cube, 144, 145
  dodecahedron, 146
  Latin cross, 155, 184
nonconvex optimization, 162
¬, not, 97
NP, Nondeterministically Polynomial, 36
NP-complete, 34, 36, 37, 58
NP-hard, 5, 17, 33–38, 40, 46, 52, 53, 56, 58, 59, 85, 86
  reduction, 38, 40, 46, 47
numbers
  constructible, 168
  integers, 1, 36, 85
  real, 1, 147

octahedron, 84
one-to-one function, 73
open problems, x, 5
  Circular Pleat, 127
  Finite Perpendiculars, 116
  Map Folding, 18
  Orthogonally Aligned Creases, 46
  P =? NP, 5, 37
  Single-Vertex, 32
  Stamp Foldings, 14
∨, OR, 37, 51
Origametry, ix, 58
origami
  algebra, 168
  axioms, 168
  design, 87–103
  history, x, 32
ORIGAMIZER, 153–166
  algorithm, 158–166
    Theorem, 161
  edge-tuck, 162
  vertex-tuck, 165
orthogonal, 4

osculating circle, 129, 130
osculating plane, 32, 130–132

P, polynomial, 36
parabola
    curvature, 129
    focus at ∞, 134, 136
    Tangents Bisection Lemma, 137
    vertex, 129, 130
parametric equations, 130, 180
perimeter, 146, 181
permutation, 11–14, 170
perpendicular, 4
perpendiculars, 109, 110, 113
    finite, 116
    wandering, 115, 179
π-calculator, 49
Pitot's Theorem, 96–99, 178
    converse, 97–99
Pitot, Henri, 96
Platonic solids, 4, 144, 155
Plus-Sign
    Lemma, 69, 127
    unit, 68–70, 174
Poisson ratio, 143
polygon, 23
    convex, 102, 103, 108, 109, 113
    nonconvex, 102, 103, 112, 113
polyhedral manifold, 153
polyhedron, 144
    convex, 120, 144
    flattening, 119
    net, 144
    nonconvex, 153
    regular, 144
polynomial
    exponent, 7, 17, 36
    growth, 6, 7, 17, 36
    time, 17, 18, 36, 38
proof, x, 5
    by contradiction, 54–56, 69
    contrapositive, 97
    rigorous, x
    sketch, x
    two-column, x
protractor, 3

quadrilateral
    four-circle, 95, 96
    tangential, 95, 118, 119

$\mathbb{R}$, real numbers, 1
race condition, 83, 86, 176
radius of gyration, 146
references, x
Resch, Ronald, 121, 132
ridge crease
    of molecule, 92
    straight skeleton, 103, 108
rigid origami, 59–86
    box, open, 60, 61
    flat folding, 85, 86
    furniture, 61
    Loop Table, 61
    solar arrays, 63
river, 99, 101, 102
robotics, soft, 144
Rule 110, 50–51, 172, 173

Schlegel diagram, 155, 156, 184
seamless, 158, 160, 161
sector, 24
self-assembly, 141, 144
self-folding, 141
    applications, 141
    cube, 144, 145
    mechanism, 141
    polyhedra, 144–147
    robots, 168
    trigger, 141
    yield, 144, 146, 147
self-unfolding, 141
\, set minus, $A \setminus B$, 85
SET PARTITION, 34–36, 171
singularity, 128
Siwanowicz, Jan, 23
Skeleton Bisection Lemma, 113
Skeleton Edge–Face Lemma, 113
Space Shuttle, 63
sphere, folding, 168
square twist, 77–83, 176
    nonrigid, 77–79
    rigid, 79–83
    tesselation, 82, 83
stamp folding, 6–18
stamps
    sheet of, 14, 15
    strip of, xi, 6, 7, 10, 32
Stanford bunny, 153–158
    metal, 166, 167
    ORIGAMIZER crease pattern, 160

star
  5-pointed, 105
  graph, 99
Steiner, Jakob, 97
stent, 141, 142
stick figure, 89, 92, 100, 101
straight skeleton, 102–103
  of convex polygon, 108
  face, 113
  of nonconvex polygon, 110
  of polyhedron, 110, 112
  structure, 112
strip algorithm, 154–158
sufficient condition, 21, 22, 26, 27, 38, 40

Tachi tubes, 133
Tachi, Tomohiro, xii, 133, 134, 158–161, 167
tesselation
  Huffman's *Arches*, 134
  Miura-unit, 76
  primal–dual, 168
  square twist, 82–83
theoretical computer science, ix, 5, 33
thick origami, 168
topological disk, 153, 159
topologically equivalent, 153
tree, 89, 177
  cut, 146, 181, 182
  embedded, 89
  leaf, 89, 100, 101
  metric, 89
TreeMaker, 87, 102, 168
triangle twist
  rigid, 83
trifold, 147–150

Turing machine, 48
  tape, 56
Turing-complete, 46–56
twist tiles, 168
Two Tangents Theorem, 95, 96, 177

Uehara, Ryuhei, ix, xii, 18
∪, union, 3

vector, 4, 71
  binormal, 128
  bit, 36
  normal, 4, 84, 128–132
  tangent, 128, 130, 132
vertex, origami, 19, 24
  bird's foot, 66
  degree, 19
  degree-4, 59–86
  degree-6, 147, 148
  flat folded, 19–32
  not flat-foldable, 19–21
vertex, straight skeleton, 109
  degree-6, 110
vesica piscis, 122, 124, 125, 132
Voronoi diagram, 166

waterbomb tube, 141–144
watertight, 158, 159, 161
wires
  noise, 46, 53, 56
  NP-hard, 40
  Turing-complete, 52
Wolfram, Stephen, 50

Zakharevich, Inna, 57

For EU product safety concerns, contact us at Calle de José Abascal, 56–1°,
28003 Madrid, Spain or eugpsr@cambridge.org.

www.ingramcontent.com/pod-product-compliance
Ingram Content Group UK Ltd.
Pitfield, Milton Keynes, MK11 3LW, UK
UKHW020051130126
466615UK00030B/112